AQA Mathematics

Unit 3 Foundation

New GCSE

WITHDRAWN

Series Editor
Paul Metcalf

Series Advisor
Andy Darbourne

Lead Author
Sandra Burns
Shaun Procter-Green
Margaret Thornton

Authors
Tony Fisher
June Haighton
Anne Haworth
Gill Hewlett
Andrew Manning
Ginette McManus
Howard Prior
David Pritchard
Dave Ridgway
Paul Winters

Published in 2010 by:
Nelson Thornes Ltd
Delta Place
27 Bath Road
CHELTENHAM
GL53 7TH
United Kingdom

11 12 13 14 / 10 9 8 7 6 5 4 3 2

A catalogue record for this book is available from the British Library

ISBN 978 1 4085 0629 5

Cover photograph: Photolibrary
Illustrations by Rupert Besley, Roger Penwill, Angela Knowles and Tech-Set Limited
Page make-up by Tech-Set Limited, Gateshead

Printed in China

Photograph acknowledgements:
Alamy: 4.7, 5.50, 15.1;
iStockphoto: 3.1, 4.1, 7.1, 8.1, 9.50, 13.1, 14.1, 14.16, 18.1, 18.18;
Fotolia: 1.5, 1.11, 2.1, 5.1, 5.11, 6.1, 6.23, C1.19, C1.22, C1.36, 9.1, 11.20, 12.1, 14.3, 14.45, 14.46, 14.47, 16.1, 17.1, C2.43;
Science Photo Library: 11.1.

Contents

1 Fractions and decimals

Objectives

Examiners would normally expect students who get these grades to be able to:

G

understand positive and negative integers

find the fraction of a shape shaded

put integers and simple fractions in order

find equivalent fractions

express simple decimals and percentages as fractions

F

add and subtract negative numbers

simplify fractions

calculate fractions of quantities

arrange fractions and decimals in order

E

multiply and divide negative numbers

express fractions as decimals and percentages

add and subtract fractions

D

find one quantity as a fraction or percentage of another

solve problems involving fractions

solve problems involving decimals

C

add and subtract mixed numbers

find the reciprocal of a number

round numbers to a given power of 10, up to three decimal places and one significant figure

multiply and divide fractions.

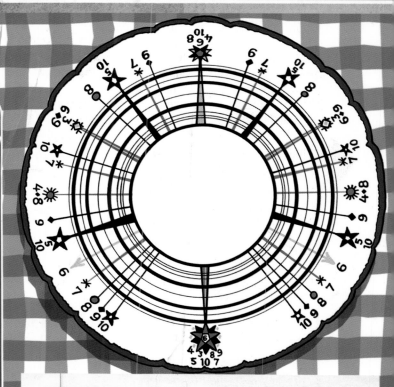

Did you know?

Plat diviseur

It is not easy to cut a pizza or a cake into pieces the same size, particularly if you need five or seven or nine pieces.

The French have a solution, the plat diviseur, or dividing plate, which has numbers marked round the edge to show where to cut for that number of pieces. Useful and beautiful!

Key terms

positive number	mixed number
negative number	rounding
integer	decimal place
equivalent fractions	significant figures
numerator	recurring decimal
denominator	reciprocal
improper fraction	

You should already know:

✔ how to add, subtract, multiply and divide simple numbers

✔ about simple fractions such as halves and quarters.

Learn... 1.1 Positive and negative numbers

Positive numbers are greater than zero and **negative numbers** are less than zero.
Zero is neither negative nor positive.
Positive and negative numbers can be **integers** (whole numbers) or non-integers such as $\frac{3}{4}$ or -1.53

A negative $(-)$ sign is always needed to show that a number is negative.
A positive sign $(+)$ can be used to show that a number is positive.

A number line helps to arrange numbers in order and to add and subtract them.

To **add** numbers, show the first number on the number line. Go to the right along the line to add a positive number or to the left to add a negative number (3 steps to the left to add -3).

The green arrows on the number line shows that $-4 + -3 = -7$

To **subtract**, show the first number on the number line. Go *left* to subtract a positive number or *right* to subtract a negative number.

The red arrow shows that $4 - -6 = 10$

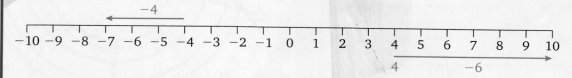

Calculating with positive and negative numbers

You need to know how to use your calculator to calculate with positive and negative numbers.

Your calculator may have a $(-)$ key. Use this to enter negative numbers into your calculator by pressing this key first, then the number.

Alternatively, you may have a $+/-$ key. To enter a negative number on your calculator, first press the number then $+/-$ to change it from positive to negative.

Example: Calculate:

     a $4 + -5$

     b -8×-6

     c $-5 + 6 + -8$

     d $60 \div -12$

Solution: **a**   Enter the calculation as $4 +$ $(-)$ 5 and press $=$
               So $4 + -5 = -1$
               (Alternatively, enter the calculation as $4 + 5$ $+/-$ and press $=$)

           b $-8 \times -6 =$ $(-)$ $8 \times$ $(-)$ $6 = 48$

           c $-5 + 6 + -8 =$ $(-)$ $5 + 6 +$ $(-)$ $8 = -7$

           d $60 \div -12 = 60 \div$ $(-)$ $12 = -5$

Example: Harry has £34.65 in his bank account and has to pay a bill for £50. How much will he have in his account when he has paid the bill?

Solution: Amount in account is £34.65 − £50 = −£15.35
              So when Harry has paid the bill, he is overdrawn: he owes the bank £15.35.

Nelson Thornes and AQA

Nelson Thornes has worked in partnership with AQA to ensure that this book and the accompanying online resources offer you the best support for your GCSE course.

All AQA endorsed resources undergo a thorough quality assurance process to ensure that their contents closely match the AQA specification. You can be confident that the content of materials branded with AQA's 'Exclusively Endorsed' logo have been written, checked and approved by AQA senior examiners, in order to achieve AQA's exclusive endorsement.

The print and online resources together unlock blended learning; this means that the links between the activities in the book and the activities online blend together to maximise your understanding of a topic and help you achieve your potential.

These online resources are available on which can be accessed via the internet at **www.kerboodle.com/live**, anytime, anywhere.

If your school or college subscribes to *kerboodle!* you will be provided with your own personal login details. Once logged in, access your course and locate the required activity.

For more information and help on how to use *kerboodle!* visit **www.kerboodle.com**.

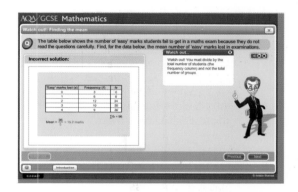

How to use this book

To help you unlock blended learning, we have referenced the activities in this book that have additional online coverage in *kerboodle!* by using this icon:

The icons in this book show you the online resources available from the start of the new specification and will always be relevant.

In addition, to keep the blend up-to-date and engaging, we review customer feedback and may add new content onto *kerboodle!* after publication.

Welcome to GCSE Mathematics

This book has been written by teachers and examiners who not only want you to get the best grade you can in your GCSE exam, but also to enjoy maths. It covers all the material you will need to know for AQA GCSE Mathematics Unit 3 Foundation. This unit allows you to use a calculator, so you will be able to use this most of the time throughout this book. Look out for calculator or non-calculator symbols (shown on the right) which tell you whether to use a calculator or not.

In the exam, you will be tested on the Assessment Objectives (AOs) below. Ask your teacher if you need help to understand what these mean.

AO1   recall and use your knowledge of the prescribed content

AO2   select and apply mathematical methods in a range of contexts

AO3   interpret and analyse problems and generate strategies to solve them.

Each chapter is made up of the following features:

Objectives

The objectives at the start of the chapter give you an idea of what you need to do to get each grade. Remember that the examiners expect you to do well at the lower grade questions on the exam paper in order to get the higher grades. So, even if you are aiming for a Grade C you will still need to do well on the Grade G questions on the exam paper.

On the first page of every chapter, there are also words that you will need to know or understand, called 'Key terms'. The box called 'You should already know' describes the maths that you will have learned before studying this chapter. There is also an interesting fact at the beginning of each chapter which tells you about maths in real life.

Learn...

The Learn sections give you the key information and examples to show how to do each topic. There are several Learn sections in each chapter.

Practise...

Questions that allow you to practise what you have just learned.

The bars that run alongside questions in the exercises show you what grade the question is aimed at. This will give you an idea of what grade you're working at. Don't forget, even if you are aiming at a Grade C, you will still need to do well on the Grades G–D questions.

These questions are harder questions to test and challenge your mathematics.

These questions are Functional Maths type questions, which show how maths can be used in real life.

These questions are problem solving questions, which will require you to think carefully about how best to answer.

These questions should be attempted **with** a calculator.

These questions should be attempted **without** using a calculator.

Assess

End of chapter questions written by examiners. Some chapters feature additional questions taken from real past papers to further your understanding.

Hint

These are tips for you to remember whilst learning the maths or answering questions.

AQA Examiner's tip

These are tips from the people who will mark your exams, giving you advice on things to remember and watch out for.

Bump up your grade

These are tips from the people who will mark your exams, giving you help on how to boost your grade, especially aimed at getting a Grade C.

Consolidation

Consolidation chapters allow you to practise what you have learned in previous chapters. The questions in these chapters can cover any of the topics you have already seen.

Practise... 1.1 Positive and negative numbers *k!* G F E D C

1 This is a table of average December temperatures in some parts of the world.

  a List the places in order of temperature, starting with the coldest.

  b How much higher is the temperature in Victoria than in Tromsø?

Place	Average temperature in degrees Celsius
Omsk, Russia	−13
Punte Arenas, Chile	+10
Scott Base, Antarctica	−5
Tromsø, Norway	−3
Victoria, Seychelles	+27

2 Work out: **a** $+12 - 15$ **c** $-17 \div -2$

  b 18×-25 **d** $25 - 30 - -18$

3 The temperature is 4°C at midday and −3°C at midnight. By how many degrees has the temperature fallen?

4 **a** A diver is at 12 m below sea level. How many metres will he have to rise to get to 2 m below sea level?

  b A diver at 3 m below sea level goes down 10 m. How far below sea level is he now?

  c A diver starts at 2 m below sea level and goes down half a metre per second. What depth is he at after 10 seconds?

5 In a quiz game, players score two points for a correct answer and lose a point for a wrong answer. If they do not answer, they score no points.

  a After 5 questions, Amy has 3 points. How many different ways could she have gained this score?

  b What possible scores could Ben have after 10 questions?

6 Becki has £45.56 in her account and then takes £100 out in cash.
How much is in her account now?

7 Samir has nothing in his bank account. He uses the account to pay a bill of £35.
How much does he have in his account now?

8 Pete's bank account is overdrawn by £36.70. He uses the account to pay a bill of £10.53. What will his account balance be now?

9 This is some information from Pip's bank statement.

Date	Details	Withdrawn	Paid in	Balance in account
17 Jul	Starting balance			− £103.25
17 Jul	Cheque number 001733	£25.00		
18 Jul	Monthly salary cheque		£1316.66	
22 Jul	Cash withdrawal	£100.00		
23 Jul	Gas bill standing order	£73.25		

Copy and complete the 'Balance in account' column to show how much Pip has in the account after each transaction.

10 Tom uses these numbers to write down five calculations with the answer −12.

$$-15 \quad -6 \quad -6 \quad -6 \quad -4 \quad -3 \quad -2 \quad 2 \quad 2 \quad 3 \quad 24$$

Write down the five calculations.

AQA *Examiner's tip*
Make sure you know how to use your calculator to do calculations with negative numbers.

Learn... 1.2 Fractions

Equivalent fractions are fractions that have the same value, such as $\frac{4}{5}$ and $\frac{8}{10}$

To find a fraction equivalent to another fraction, multiply or divide the **numerator** and the **denominator** of the fraction by the same number.

$$\overset{\times 2}{\frac{1}{2} \underset{\times 2}{=} \frac{2}{4}} \qquad \overset{\times 6}{\frac{1}{2} \underset{\times 6}{=} \frac{6}{12}} \qquad \overset{\div 6}{\frac{6}{12} \underset{\div 6}{=} \frac{1}{2}}$$

Fractions should be given in their simplest form when possible.
The simplest fraction in the list $\frac{1}{2}, \frac{2}{4}, \frac{3}{6}$ and $\frac{6}{12}$ is $\frac{1}{2}$

Your calculator can simplify a fraction such as $\frac{8}{10}$ using the fraction key ▣ or $\boxed{a\frac{b}{c}}$

Press ▣ 8 ▶ 10 = $\frac{4}{5}$ (Alternatively, 8 $\boxed{a\frac{b}{c}}$ 10 = $\frac{4}{5}$)

If you enter a top-heavy fraction (also known as an **improper fraction**), the calculator will simplify it and may give it as a **mixed number**. Find out what your calculator will do. Some calculators will change an improper fraction to a mixed number (and vice versa) if you press SHIFT S⇔D.

If your calculator will not change between improper fractions and mixed numbers, you will have to do it yourself.

To change an improper fraction such as $\frac{15}{4}$ to a mixed number, find how many times 4 goes into 15. It goes 3 times with 3 left over, so $\frac{15}{4}$ is $3\frac{3}{4}$ ($\frac{15}{4}$ is 15 quarters, 12 quarters make 3 whole ones and there are 3 quarters left over.)

To change a mixed number such as $3\frac{3}{4}$ to an improper fraction, multiply 3 by 4 to give 12 and add on 3, to give $\frac{15}{4}$. (3 whole ones are 12 quarters and there are 3 more quarters to be added on, giving 15 quarters.)

AQA Examiner's tip

Find out how to use your calculator to simplify fractions and to change between mixed numbers and top-heavy fractions.

Example: Write each fraction in its simplest form.

    a $\frac{10}{15}$

    b $\frac{45}{100}$

    c $\frac{175}{50}$

Solution:

    a ▣ 10 ▶ 15 = $\frac{2}{3}$ (or 10 $\boxed{a\frac{b}{c}}$ 15 = $\frac{2}{3}$)

    b ▣ 45 ▶ 100 = $\frac{9}{20}$ (or 45 $\boxed{a\frac{b}{c}}$ 100 = $\frac{9}{20}$)

    c ▣ 175 ▶ 50 = $\frac{7}{2}$ = $3\frac{1}{2}$ (or 175 $\boxed{a\frac{b}{c}}$ 50 = $3\frac{1}{2}$)

Your calculator can also do fraction calculations. Use the fraction key to enter the fractions and use the normal ×, ÷, − and + keys to do the calculations.

Example: Calculate:

    a $\frac{2}{3} + \frac{4}{5}$

    b $1\frac{3}{4} \times 2\frac{1}{2}$

    c $\frac{3}{8} \div \frac{7}{12}$

Solution:

    a ▣ 2 ▶ 3 ▶ + ▣ 4 ▶ 5 = $\frac{22}{15}$ (or 2 $\boxed{a\frac{b}{c}}$ 3 + 4 $\boxed{a\frac{b}{c}}$ 5 = $\frac{22}{15}$)

    b Shift ▣ 1 ▶ 3 ▶ 4 ▶ × Shift ▣ 2 ▶ 1 ▶ 2 = $\frac{35}{8}$ SHIFT S⇔D = $4\frac{3}{8}$

       (or 1 $\boxed{a\frac{b}{c}}$ 3 $\boxed{a\frac{b}{c}}$ 4 × 2 $\boxed{a\frac{b}{c}}$ 1 $\boxed{a\frac{b}{c}}$ 2 = $4\frac{3}{8}$)

    c ▣ 3 ▶ 8 ▶ ÷ ▣ 7 ▶ 12 = $\frac{9}{14}$ (or 3 $\boxed{a\frac{b}{c}}$ 8 ÷ 7 $\boxed{a\frac{b}{c}}$ 12 = $\frac{9}{14}$)

To work out one quantity as a fraction of another, make the first quantity the numerator of the fraction and the second the denominator. You need to make sure the units of the two quantities are the same before you do this. Use your calculator to simplify the fraction.

So 20 minutes as a fraction of one hour = 20 minutes as a fraction of 60 minutes = $\frac{20}{60} = \frac{1}{3}$

Example: Write 30p as a fraction of £6.

Solution: 30p as a fraction of £6 is 30p as a fraction of 600p, which is $\frac{30}{600}$

Use your calculator to simplify the fraction. 🖩 30 ▶ 600 ▶ = $\frac{1}{20}$ (or 30 $\boxed{a\frac{b}{c}}$ 600 = $\frac{1}{20}$)

Example: In a mathematics exam, there are 35 marks for Section A and 45 marks for Section B. What fraction of the total exam marks are given for Section A?

Solution: Total number of marks for the exam is 35 + 45 = 80

Fraction of marks for Section A is $\frac{35}{80} = \frac{7}{16}$
(Simplify the fraction on your calculator.)

AQA Examiner's tip
Remember to give your answers in their simplest fraction form.

Practise... 1.2 Fractions

G F E D C

1 **a** Use your calculator to simplify these fractions.
    i $\frac{12}{60}$ **ii** $\frac{22}{33}$ **iii** $\frac{75}{100}$ **iv** $\frac{125}{200}$ **v** $\frac{45}{180}$

    b Simplify these fractions to find the odd one out.
    i $\frac{7}{21}$ **ii** $\frac{15}{45}$ **iii** $\frac{1}{3}$ **iv** $\frac{9}{36}$ **v** $\frac{12}{36}$

2 $\frac{14}{16}$ is equivalent to $\frac{7}{8}$. Write down two more fractions equivalent to $\frac{7}{8}$

3 Use your calculator for this question.

    a Express these improper fractions as mixed numbers.
    i $\frac{10}{7}$ **ii** $\frac{15}{2}$ **iii** $\frac{17}{4}$ **iv** $\frac{105}{10}$ **v** $\frac{27}{5}$

    b Express these mixed numbers as improper fractions.
    i $2\frac{3}{4}$ **ii** $3\frac{2}{3}$ **iii** $1\frac{7}{8}$ **iv** $10\frac{1}{2}$ **v** $5\frac{5}{6}$

4 Use your calculator to write the first quantity as a fraction of the second.

    a 25 minutes, $2\frac{1}{2}$ hours   **d** 20 cm, 1.8 m

    b Half an hour, three hours   **e** 400 g, 1.2 kg

    c 40p, £2.40

5 Here are nine shapes.

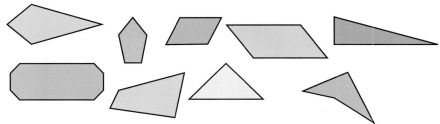

    a How many quadrilaterals are there?

    b What fraction of the total number of shapes are quadrilaterals?

    c What fraction of the total number of shapes are parallelograms?

Hint
A quadrilateral is a shape with 4 sides.

D

6   Pigs have an average weight of 150 pounds and a stomach weighing 6 pounds.

 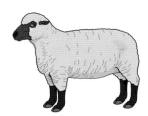

   a   What fraction of a pig's total weight is the weight of its stomach?

Sheep have a different digestive system. They have an average weight of 120 pounds and a stomach weighing 30 pounds.

   b   What fraction of a sheep's total weight is the weight of its stomach?

⚠ 7   Which of these expressions can be used to find x minutes as a fraction of y hours?

$$\frac{60x}{y} \qquad x \div 60y \qquad 60xy \qquad \frac{x}{60y} \qquad \frac{60y}{x}$$

⚠ 8   Find the missing fractions.

   a $\frac{2}{3} \times \square = 1$ **c** $\frac{2}{3} - \square = 1$ **e** $\frac{p}{q} \times \square = 1$ **g** $\frac{p}{q} - \square = 1$

   b $\frac{2}{3} + \square = 1$ **d** $\frac{2}{3} \div \square = 1$ **f** $\frac{p}{q} + \square = 1$ **h** $\frac{p}{q} \div \square = 1$

⚙ 9   Sue's recipe for custard needs $1\frac{1}{2}$ pints of milk and $\frac{3}{8}$ pint of cream. How many pints is this altogether?

⚙ 10   A pair of shoes costing £35 is reduced by £3.50 in a sale.

   a   What fraction of the original price is the reduction?

   b   What fraction is the sale price of the original price?

⚙ 11   In New York on 6 September 2009, the sunrise time was 06:27 and the sunset time was 19:21. What fraction of the time from midnight on 5 September to midnight on 6 September was daylight?

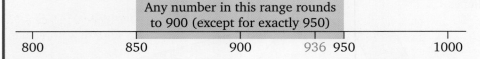

Learn...   **1.3 Working with fractions and decimals**

Rounding

Numbers can be **rounded** to make them easy to work with and to understand. For example, if parents want to know how many students there are in a school, they probably want to know the approximate number, 900 for example, rather than the exact number 936.

Rounded to the nearest 100, 936 is 900. It is between 900 and 1000 and nearer to 900.

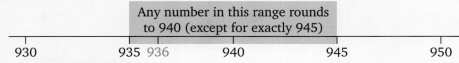

Any number in this range rounds to 900 (except for exactly 950)

800   850   900   936   950   1000

Rounded to the nearest 10, 936 is 940. It is between 930 and 940 and nearer to 940.

Any number in this range rounds to 940 (except for exactly 945)

930   935   936   940   945   950

You also need to be able to round decimal numbers to the nearest integer, to one **decimal place**, two decimal places and so on.

Example: Round the number 7.846 to:

    a the nearest integer   **b** one decimal place   **c** two decimal places.

Solution: **a** 7.846 is between 7 and 8 and nearer to 8.

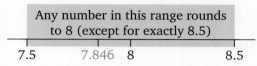

    So, to the nearest integer, 7.846 is 8.

    b 7.846 is between 7.8 and 7.9 and nearer to 7.8. So 7.846 to one decimal place is 7.8

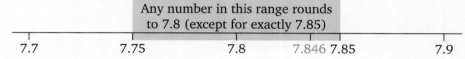

    c To two decimal places, 7.846 is 7.85; it is between 7.84 and 7.85 and nearer to 7.85

Significant figures

You can also round numbers to different numbers of **significant figures**.

936 rounds to 900 and 0.0936 rounds to 0.09. The first is rounded to the nearest 100 and the second to two decimal places, but they are both rounded to one significant figure. There is one figure, 9, in each that is most important when considering the size of the number. This number is the first (or most) significant figure. Both numbers also need zeros to show you the place value of the 9 but these zeros are not significant figures.

(Zeros **can** be significant; for example, 1023 rounded to the nearest 10 is 1020. The zero in the hundreds position is significant but the zero in the units position is not.)

To two significant figures, 936 is 940 and
0.0936 is 0.094

> **AQA** *Examiner's tip*
>
> Make sure you understand the difference between rounding to a number of significant figures and rounding to a number of decimal places.

Example: Round these numbers to:

    a the nearest integer   **i** 44.79   **ii** 0.5678   **iii** 204.45   **iv** 0.0235

    b one decimal place   **i** 44.79   **ii** 0.5678   **iii** 204.45   **iv** 0.0235

    c one significant figure.   **i** 44.79   **ii** 0.5678   **iii** 204.45   **iv** 0.0235

Solution: **a** **i** 44.79 is between 44 and 45 and nearer to 45. So 44.79 to the nearest integer is 45

        ii 0.5678 is between 0 and 1 and nearer to 1. So 0.5678 to the nearest integer is 1

        iii 204.45 is between 204 and 205 and nearer to 204. So 204.45 to the nearest integer is 204

        iv 0.0235 is between 0 and 1 and nearer to 0. So 0.0235 to the nearest integer is 0

    b **i** 44.79 is between 44.7 and 44.8 and nearer to 44.8. So 44.79 to one decimal place is 44.8

        ii 0.5678 is between 0.5 and 0.6 and nearer to 0.6. So 0.5678 to one decimal place is 0.6

        iii 204.45 is between 204.4 and 204.5 and nearer to 204.5. So 204.45 to one decimal place is 204.5

        iv 0.0235 is between 0.0 and 0.1 and nearer to 0.0. So 0.0235 to one decimal place is 0.0

    c **i** 44.79 is between 40 and 50 and nearer to 40. So 44.79 to one significant figure is 40.

        ii 0.5678 is between 0.5 and 0.6 and nearer to 0.6. So 0.5678 to one significant figure is 0.6

        iii 204.45 is between 200 and 300 and nearer to 200. So 204.45 to one significant figure is 200

        iv 0.0235 is between 0.02 and 0.03 and nearer to 0.02. So 0.0235 to one significant figure is 0.02

Expressing fractions as decimals

Any fraction can be expressed as a decimal.

Some simple examples that you may already know are $\frac{1}{2} = 0.5$, $\frac{3}{4} = 0.75$, $\frac{3}{10} = 0.3$

To express a fraction as a decimal, use your calculator to divide the numerator by the denominator.

$\frac{3}{4} = 3 \div 4 = 0.75$ (Make sure your calculator is set to give a decimal answer not a fraction answer.)

0.75 is a **terminating decimal**.

$3 \div 4$ works out exactly, the calculation comes to an end after two decimal places.

Some fractions become **recurring decimals**. The division calculation does not stop.
Instead one digit, or a group of digits, repeats forever.

$\frac{1}{3} = 1 \div 3 = 0.333...$ This can also be written as $0.\dot{3}$

$\frac{2}{3} = 2 \div 3 = 0.666...$ This can also be written as $0.\dot{6}$

The calculator display shows that $\frac{2}{3}$ is 0.6666666667. The calculator can show only a limited number of digits and it rounds the final one in the display.

When decimals are used in calculations you have to round them appropriately.

$\frac{2}{3}$ to one decimal place is 0.7

$\frac{2}{3}$ to two decimal places is 0.67

$\frac{2}{3}$ to three decimal places is 0.667

$\frac{2}{3}$ to four decimal places is 0.6667 and so on.

Expressing fractions as percentages

To express a fraction as a percentage, change the fraction to a decimal and then multiply the decimal by 100%

$\frac{3}{4} = 0.75$

$0.75 \times 100\% = 75\%$

> **AQA Examiner's tip**
>
> In calculations, use all the figures already in your calculator whenever possible. Do not re-enter the number unless you really have to.

Example: **a** What is $\frac{5}{12}$ as a decimal? Write your answer correct to three decimal places.

  b Use your answer to write $\frac{5}{12}$ as a percentage.

Solution: **a** $\frac{5}{12} = 5 \div 12 = 0.41666...$
      $= 0.417$ to three decimal places.

> **AQA Examiner's tip**
>
> Make sure you always divide the numerator by the denominator when changing a fraction to a decimal.

  b So $\frac{5}{12}$ as a percentage is 41.7% (3 sf) or 42%, correct to the nearest whole percent.

Arranging fractions in order

To arrange fractions in order of size, first change them to decimals or percentages.

It is not easy to see which fraction, $\frac{3}{4}$ or $\frac{7}{9}$, is bigger, so change them both to decimals.

$\frac{3}{4} = 0.75$ and $\frac{7}{9} = 0.\dot{7}$

You can see that $\frac{7}{9}$ is a bit bigger than $\frac{3}{4}$

Example: Arrange these fractions in order of size, starting with the smallest.

  $\frac{2}{3}, \frac{3}{5}, \frac{11}{18}, \frac{7}{11}$

Solution: $\frac{2}{3} = 0.\dot{6}$ $\frac{3}{5} = 0.6$, $\frac{11}{18} = 0.61\dot{1}$, $\frac{7}{11} = 0.\dot{6}\dot{3}$ $0.\dot{6}\dot{3}$ means 0.636363...

  So the fractions in order of size are $\frac{3}{5}, \frac{11}{18}, \frac{7}{11}, \frac{2}{3}$

  Alternatively, change them to percentages.

  $\frac{2}{3} = 0.\dot{6} = 67\%$, $\frac{3}{5} = 0.6 = 60\%$, $\frac{11}{18} = 0.61\dot{1} = 61\%$, $\frac{7}{11} = 0.\dot{6}\dot{3} = 64\%$

  The percentages here have been rounded to two significant figures.

  Once again, the fractions in order of size are $\frac{3}{5}, \frac{11}{18}, \frac{7}{11}, \frac{2}{3}$

Reciprocals

When two numbers multiply together to make 1, the numbers are called the **reciprocals** of each other.

So 2 and $\frac{1}{2}$ are the reciprocals of each other because $2 \times \frac{1}{2} = 1$

Your calculator can work out reciprocals both as fractions and as decimals. Look for the button labelled $\frac{1}{x}$ or x^{-1} and find out how to use it.

Bump up your grade

To get a Grade C you need to know how to work out reciprocals.

Example: Find the reciprocals of 0.2, $\frac{9}{20}$, 0.89

Solution: Reciprocal of 0.2 is 0.2 $\boxed{x^{-1}}$ $\boxed{=}$ 5

Reciprocal of $\frac{9}{20}$ is $\frac{20}{9} = 2\frac{2}{9}$

Reciprocal of 0.89 is 0.89 $\boxed{x^{-1}}$ $\boxed{=}$ 1.12 to three significant figures.

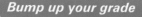 **Examiner's tip**

To find the reciprocal of a **fraction**, swap the numerator and denominator.

To find the reciprocal of a **mixed number**, convert it to an improper (top-heavy) fraction first. Then swap the numerator and denominator.

Fractions of quantities

You can work out a fraction of a given quantity on your calculator.

Example: Find four fifths of £3.50.

Solution: Four fifths of £3.50 is $\frac{4}{5} \times$ £3.50 = £2.80

Example: A jacket costing £22 is reduced by $\frac{1}{3}$ in a sale. What is the sale price?

Solution: $\frac{1}{3}$ of £22 = $\frac{1}{3} \times$ £22 = £7.33 (rounded to the nearest penny).

So the sale price is £22 − £7.33 = £14.67

(You could do this in one step by working out $\frac{2}{3}$ of £22 rather than working out $\frac{1}{3}$ and subtracting it.)

1.3 Working with fractions and decimals

Practise...

G F E D C

1 There are approximately one million dairy cows in the US. $\frac{4}{5}$ of these are Holsteins. Approximately how many Holsteins are there in the US?

2 A pizza is cut into five equal pieces as shown. Work out the size of the angle marked x.

3 **a** Calculate $\frac{3}{4}$ of:

i 56   **ii** £105   **iii** £15.65   **iv** 3.5 metres   **v** 500 grams

b Calculate $\frac{3}{8}$ of:

i 254   **ii** £25   **iii** $15   **iv** 3 hours   **v** 1 km

Hint

Round your answers if they do not work out exactly.

F

E

4 Express these fractions as:

    a decimals   **b** percentages.

    i $\frac{4}{5}$ **ii** $\frac{9}{10}$ **iii** $\frac{11}{20}$ **iv** $\frac{23}{50}$ **v** $\frac{67}{100}$

5 Which of these fractions are equivalent to recurring decimals?

    a $\frac{3}{5}$ **b** $\frac{9}{11}$ **c** $\frac{5}{6}$ **d** $\frac{7}{20}$ **e** $\frac{4}{15}$

C

6 Round these numbers to:

    a the nearest 10

    b the nearest integer

    c one decimal place

    d two decimal places

    e three decimal places

    f one significant figure.

    i 12.89   **ii** 54.5   **iii** 109.87   **iv** 4.756   **v** 0.836

7 Round these numbers and quantities to one significant figure.

    a The height of Mount Everest, 8848 m

    b The length of the River Nile, 4135 miles

    c The number of words in the Bible, 181 253

    d The estimated population of the UK in 2013, which is 63 498 000 people

    e The diameter of a pound coin, 2.250 cm

    f The weight of a wren, 0.026 kg

8 **a** Match each number with its reciprocal.

$\frac{5}{9}$	1
1	$\frac{1}{7}$
0.4	2.5
15	$1\frac{4}{5}$
7	0.0667

    b Which number is the reciprocal of itself?

    c What happens when you multiply a number by its reciprocal?

    d Use your calculator to try to find the reciprocal of zero. What happens?

9 **a** Multiply 150 by $\frac{1}{4}$. What do you have to multiply the answer by to get back to 150?

    b Choose another number. Multiply it by 2.5.
What do you have to multiply the answer by to get back to the original number?

    c What do your answers to part **a** and part **b** tell you about reciprocals?

10 Match each number with its reciprocal.

$\frac{1}{x}$	$\frac{d}{c}$
$\frac{c}{d}$	$0.1x$
$\frac{10}{x}$	x
$15x$	$\frac{5}{2}x$
$0.4x$	$\frac{1}{15x}$

11 The perimeter of a square is 20 cm.
The sides of the square are enlarged by 10%

    a   By what percentage is the perimeter of the enlarged square bigger than that of the original?

    b   By what percentage is the area of the enlarged square bigger than that of the original?

12 Here are the ingredients for crème brulée pudding for eight people.

> 8 egg yolks
>
> 1 litre (1000 ml) cream
>
> 225 g sugar

Jane is making crème brulée for two people.
Work out how much of each ingredient she will need.

13 Jake scored 12 marks out of 25 in his first mathematics test and 14 out of 30 in his second.
In which test did he do better? Show how you found your answer.

14 Paul uses the standard formula for converting Fahrenheit temperatures (°F) to Celsius temperatures (°C),

$$C = \frac{5}{9}(F - 32)$$

Tim uses another version, $C = 0.56(F - 32)$

Find the difference between Paul's answers and Tim's answers when converting:

    a   an oven temperature of 400°F to Celsius

    b   a room temperature of 20°F to Celsius.

15 $\frac{5}{8}$ of a number is 40. What is the number?

16 $\frac{3}{4}$ of $\frac{2}{5}$ of a number is 18. What is the number?

Assess ⓚ!

1 Which of these fractions is not equivalent to the other three?

$\frac{5}{6}, \frac{10}{12}, \frac{14}{18}, \frac{20}{24}$

Show how you worked out your answer.

2 Paul has £153.29 in his bank account. His account has an overdraft.
He writes a cheque for £209.25. How much will he have in his account after the
cheque goes through?

3 Find the missing numbers.

    a $2.5 \times \square = -5$ **c** $-4.5 + \square = -2$

    b $\frac{1}{2} \times \square = 1$ **d** $\square \times -3 = 1.5$

4 Which is the biggest fraction in this list?

    a $\frac{8}{9}$ **b** $\frac{6}{7}$ **c** $\frac{5}{6}$ **d** $\frac{9}{11}$ **e** $\frac{4}{5}$

G

F

E

5 Match each fraction with its percentage.

$\frac{2}{3}$	
$\frac{17}{20}$	
$\frac{3}{8}$	
$\frac{5}{9}$	
$\frac{4}{5}$	

56%
38%
67%
85%
80%

D

6 Jyoti has a dozen eggs. She uses three eggs for breakfast and two in a cake. What fraction of her eggs does she have left?

7 One day John gets 36 e-mails. 28 of them are work related and the rest are personal e-mails. What fraction of his e-mails are work related?

8 A video game runs at 30 frames a second. What is the total time taken by three scenes of 33 frames, 44 frames and 32 frames?

9 A skirt needs $1\frac{3}{4}$ yards of fabric and a jacket needs $2\frac{1}{8}$ yards. How much fabric is needed altogether for four skirts and three jackets?

10 In a clinical trial of two new drugs, 2135 out of 3000 patients taking Drug A got better and 1855 out of 2500 patients taking Drug B got better. Which drug appears to be the more effective?

11 A number when rounded to three decimal places is 0.015. Write down three possible values of the number. What is its smallest possible value?

12 Drew's annual salary is increased by $\frac{1}{4}$ and becomes £36 400. What was his original salary?

13 There are three girls and seven boys in the chess club. One more boy and one more girl join the club. Is the percentage of girls in the club now more, less or the same? Show how you worked out your answer.

C

14 Which of these numbers are the reciprocal of $\frac{5}{8}$?

  a 0.625   **c** $1\frac{3}{8}$ **e** $\frac{5}{8}$

  b $\frac{8}{5}$ **d** $1\frac{3}{5}$ **f** 1.6

15 The price of a shirt is reduced by one-fifth in a sale. The next week, the sale price is increased by one-fifth.

Is the final price less than, greater than or the same as the original price? Explain how you worked out your answer.

AQA Examination-style questions 🄺

1 A supermarket sells 600 kg of potatoes.
$\frac{1}{2}$ of the potatoes are sold in 10 kg bags.
$\frac{3}{10}$ of the potatoes are sold in 5 kg bags.
The rest of the potatoes are sold in 3 kg bags.
How many 3 kg bags does the supermarket sell?

(5 marks)

AQA 2006

2 Angles

Examiners would normally expect students who get these grades to be able to:

G

recognise acute, obtuse and right angles

understand the terms 'perpendicular' and 'parallel'

identify scalene, isosceles, equilateral and right-angled triangles

F

recognise reflex angles

estimate angles and measure them accurately

use properties of angles at a point and on a straight line

E

use angle properties of triangles including the sum to 180°

show that the exterior angle of a triangle is equal to the sum of the interior opposite angles

D

recognise corresponding, alternate and interior angles in parallel lines

understand and use three-figure bearings.

Did you know?

Angles in snooker

If you want to play snooker well, it is important to understand how the balls rebound from the cushions around the edge of the table. World class snooker players improve their game by spending many hours practising how to make the balls rebound at the correct angles. Use of angles is particularly good for making sure you are not 'snookered'!

Key terms

acute angle
right angle
obtuse angle
straight angle
reflex angle
vertically opposite angles
perpendicular
parallel
transversal
alternate angles

corresponding angles
interior
bearing
triangle
equilateral triangle
isosceles triangle
scalene triangle
right-angled triangle
exterior angle

You should already know:

✔ how to recognise and use simple fractions

✔ how to use a protractor

✔ compass directions: north, south, east, west

✔ the meaning of clockwise and anticlockwise.

Learn... 2.1 Angles

An angle is an amount of turn. It is usually measured in degrees.

One full turn, or complete circle, measures 360°.

One half turn measures 180°.

An angle between 0° and 90° is an **acute angle**.

An angle of exactly 90° is a **right angle**.
It is always marked with a small square.

An angle of between 90° and 180° is an **obtuse angle**.

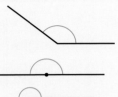

An angle of exactly 180° is called a **straight angle**.

An angle between 180° and 360° is a **reflex angle**.

Example: State whether each of the following angles is acute, obtuse or reflex.

  a   256°   **b**   79°   **c**   112°   **d**   91°

Solution: **a**   256° is between 180° and 360° so it is a reflex angle.

  b   79° is between 0° and 90° so it is an acute angle.

  c   112° is between 90° and 180° so it is an obtuse angle.

  d   91° is between 90° and 180° so it is an obtuse angle.

Practise... 2.1 Angles

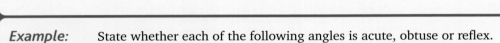   G F E D C

1   How many degrees are there in:

  a   a quarter turn

  b   one-third of a turn

  c   one-sixth of a turn?

2   What fraction of a full turn is:

  a   30°   **b**   45°   **c**   270°   **d**   180°?

3   The diagram shows the four main compass directions:
north, south, east and west.

  a   Clare faces north and makes a half turn.
    Which way is she facing now?

  b   Jane faces east and makes a quarter turn clockwise.
    Which way is she facing now?

  c   Leroy faces west and makes a quarter turn anticlockwise.
    Which way is he facing now?

  d   Gemma has made a half turn and is now facing west.
    Where was she facing before she turned?

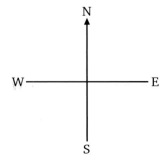

4 State whether each of the following angles is acute, obtuse or reflex. Give a reason for each answer.

a	125°	**e**	157°	**i**	148°	**m**	300°
b	62°	**f**	46°	**j**	206°	**n**	25°
c	89°	**g**	195°	**k**	100°	**o**	98°
d	312°	**h**	10°	**l**	200°		

5 For each marked angle:

i write down whether it is an acute angle, an obtuse angle or a reflex angle

ii **estimate** the size of the angle

iii use a protractor to **measure** the size of the angle.

Check that your answers agree with your answers to part **ii**.

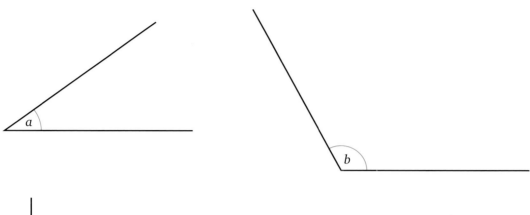

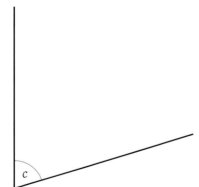

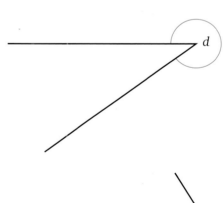

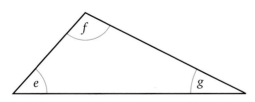

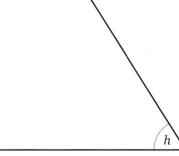

Hint
You can draw in 90° lines on a copy of the angle or divide 90° in two to help you estimate the size of the angle.

G
F

Learn... 2.2 Angles and lines

Angles meeting at a point form a complete turn or circle.
So angles at a point add up to 360°.

Angles on a straight line form a half turn.
So angles on a straight line add up to 180°.

Angles on a straight line

$p + q + r = 180°$

Where two lines cross, the **vertically opposite angles** are equal.

Angles at a point

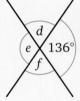

$a + b + c + d = 360°$

Two lines crossing or meeting at right angles are called **perpendicular** lines.

Example: Work out the size of the marked angles.

a

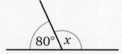

Not drawn accurately

b

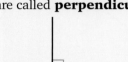

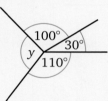

c

Solution:

a Angles on a straight line add up to 180°.
$$80° + x = 180°$$
So $\quad x = 180° - 80° = 100°$

b Angles at a point add up to 360°.
So $\quad y = 360° - (110° + 30° + 100°)$
$$= 360° - 240°$$
$$y = 120°$$

c $d = 180° - 136° = 44°$ (angles on a straight line)
$e = 136°$ (vertically opposite angles)
$f = 44°$ (vertically opposite to d)

Practise... 2.2 Angles and lines

G F E D C

F

1 Calculate the size of the angles marked with letters.

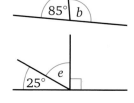

Not drawn accurately

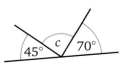

2 Calculate the size of the angles marked with letters.

Not drawn accurately

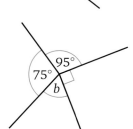

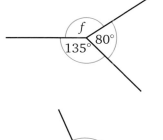

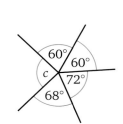

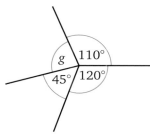

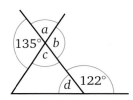

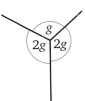

3 Calculate the size of the angles marked with letters *a–h*.

Not drawn accurately

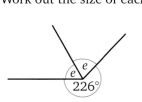

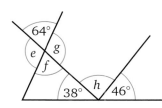

▲ 4 Work out the size of each of the angles marked with letters.

AQA *Examiner's tip*

Angles marked with the same letter in a diagram will be equal.

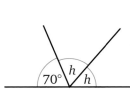

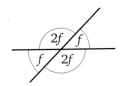

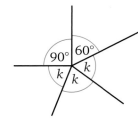

Not drawn accurately

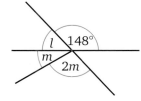

5 Two angles on a straight line are in the ratio 3 : 2

Work out the two angles.

6 Three angles meet at a point. The smallest is 10° smaller than the middle-sized angle. The largest is 10° larger than the middle-sized angle.

Work out the three angles.

Learn... 2.3 Angles and parallel lines

Two lines which stay the same perpendicular distance apart are called **parallel** lines.

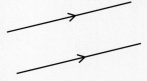

The arrows show that the lines are parallel.

A line through the two parallel lines is called a **transversal**.

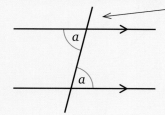   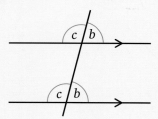

Several pairs of equal angles are formed between the parallel lines and the transversal.

The angles marked a are **alternate angles**. They are equal.

The angles are on opposite sides of the transversal.

The angles marked b are **corresponding angles**. They are equal.

The angles are in similar positions on the same side of the transversal.

The angles marked c are another pair of corresponding angles.

Angles d and e are **interior** or **allied angles**.

They always add up to 180°.

So $d + e = 180°$.

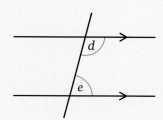

Example: Work out the value of the marked angles. Give reasons for your answers.

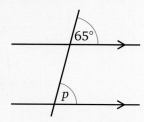   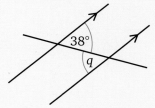   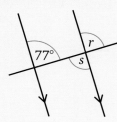

Solution: $p = 65°$ (corresponding angles)

$q = 38°$ (alternate angles)

$r = 77°$ (corresponding angles)

$s = 77°$ (alternate angles or vertically opposite angles)

Notice that it is often possible to find angles by more than one method (as for angle s).

AQA **Examiner's tip**

Always give the correct term, such as corresponding angles, when you are asked to give the reasons for your answers.

Practise...   2.3 Angles and parallel lines   D

In this exercise the diagrams are not drawn accurately.

1   Work out the values of angles x, y and z.

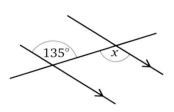   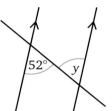   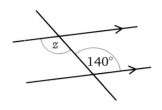

2   Work out the values of the angles marked with letters.
Give reasons for your answers.

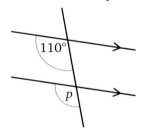   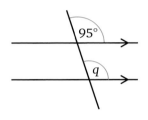   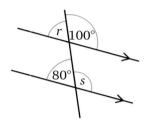

3   Work out the values of angles x, y and z.
Give reasons for your answers.

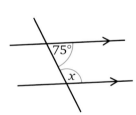   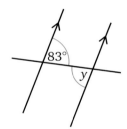   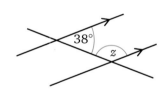

4   Work out the values of the angles marked with letters.
Give reasons for your answers.

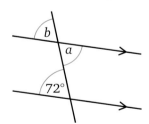   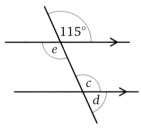   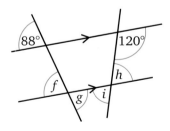

5   **a**   Are lines AB and CD parallel?   **b**   Are lines EF and GH parallel?
Give a reason for your answer.   Give a reason for your answer.

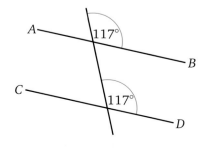   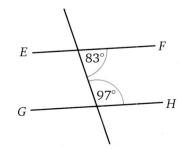

Hint

Remember you can use facts about angles at a point or on a line as well as angle properties of parallel lines.

6 Calculate the size of the angles marked with letters.

a

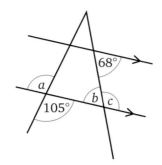

c

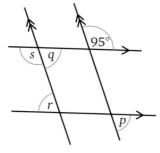

b

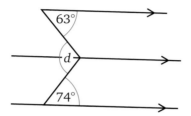

d

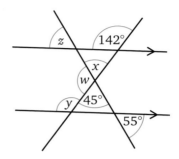

Learn... 2.4 Bearings

Directions can be described using the points of the compass such as south, north, west and so on.

Directions can also be described using three-figure **bearings**.

A three-figure bearing gives the angle measured in a clockwise direction from the **north** line.

Angles of less than 100° need a zero placed in front to make them three figures.
For example, a bearing of 80° is written as 080°.

So **south** has a bearing of 180°, **east** has a bearing of 090° and **west** has a bearing of 270°.

Other directions can all be described using bearings.

Example: Write down the bearings of the following compass directions.

   a   north-east   **b**   south-west

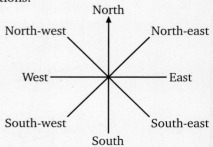

Solution: **a**   North-east is halfway between north and east
so the bearing is halfway between 0° and 090°.
So north-east is on a bearing of 045°.

   b   South-west is halfway between south and west.
Halfway between 180° and 270° is 225°,
so south-west is on a bearing of 225°.

Example: What are the three-figure bearings of directions A, B and C from the point O?

a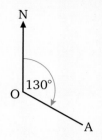

b

c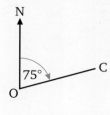

Solution: **a** Bearing of 130° **b** Bearing of 325° **c** Bearing of 075°

Bearings from one place to another can be found by measurement or by calculation.

Example: What are the three-figure bearings of directions A from B, P from Q, and D from E?

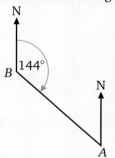

Solution: The bearing of *A* **from** *B* The bearing of *P* **from** *Q* The bearing of *D* **from** *E*
is 144°   is 055°   is 250°

Example: Branton is on a bearing of 105° from Averby.

Work out the bearing of Averby from Branton.

Solution: Draw a sketch.

If you walk from Averby to Branton you are walking
on a bearing of 105°.

To return to Averby from Branton you have to turn
round to face the opposite direction (half a turn
or 180°).

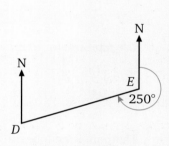

So in total you are now on a bearing of 105° + 180°
from north.

So facing towards Averby **from** Branton, the angle measured clockwise from north is
105° + 180° = 285°.

So the bearing from Averby **from** Branton is 285°.

> **Hint**
>
> Remember to put the north line at the place you are working
> the bearing out '**from**'.
>
> E.g. in this example the bearing is 'from Branton' so the
> north line is drawn at Branton then the angle is measured
> clockwise from north around to the line joining Branton to
> Averby (as shown by the arrow on the angle arc).

Practise... 2.4 Bearings

D

1 For each diagram, write down the three-figure bearing of D from E.

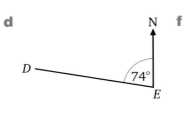 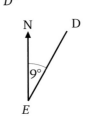

a N
E 115°
D

c N
D ⌐ E

e N
132° E
D

b N
D
29°
E

d N
D
74°
E

f N D
9°
E

2 Use a protractor to draw accurate diagrams to represent these bearings.

   a 140°   **c** 210°   **e** 085°   **g** 163°

   b 045°   **d** 320°   **f** 108°   **h** 258°

3 Thatham is on a bearing of 078° from Benton.

Work out the bearing of Benton from Thatham.

Use a sketch to help you.

4 Newby is on a bearing of 250° from Reddington.

Work out the bearing of Reddington from Newby.

 5 Here is a map of an island.

P is a port, T is a town and B is a beach. H_1 and H_2 are hotels.

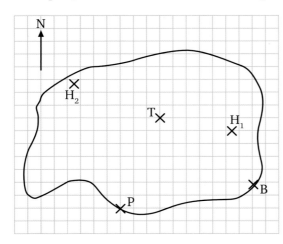

   a Which hotel is on a bearing of 055° from the port?

   b Measure and write down the bearing of the beach from the port.

   c Measure and write down the bearing of the port from the town.

Learn...   **2.5 Angles and triangles**

There are three main types of **triangle**:

| **Equilateral triangle** | **Isosceles triangle** | **Scalene triangle** |

                    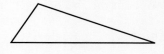

- Three equal sides
- Three equal angles
- Each angle $= \dfrac{180°}{3} = 60°$

- Two equal sides
- Two equal angles
- The equal angles are opposite the equal sides and are called the base angles even when the triangle is on its side.

- No equal sides
- No equal angles

The same number of dashes on the sides show equal sides and the same number of arcs on angles show equal angles.

Any triangle with a right angle is called a **right-angled triangle**. A right-angled triangle can be scalene or isosceles.

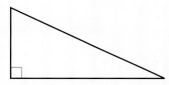   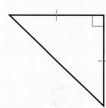

right-angled scalene triangle   right-angled isosceles triangle

The sum of the three angles in any triangle is always 180°. You can prove this as follows.

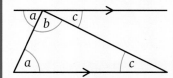   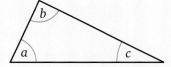

The alternate angles are equal. Therefore $a + b + c = 180°$

If a side of the triangle is extended an **exterior angle** is formed.

The exterior angle of a triangle is always equal to the sum of the two interior opposite angles.

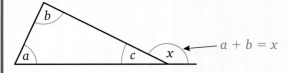

This can be shown to be true using angle properties you have learnt previously.

If the third angle inside the triangle is c,

then $a + b + c = 180°$ (angle sum of a triangle)

and $c + x = 180°$ (angles on a straight line)

So $a + b = x$

Example:

In a triangle *ABC*, angle *BAC* = 50° and angle *BCA* = 70°

Draw a sketch of the triangle.

Work out the size of angle *ABC*.

Solution: Angle *BAC* is the angle between sides *BA* and *AC*, that is the angle inside the triangle at *A*.

Angle *BCA* is the angle between sides *BC* and *CA*, that is the angle inside the triangle at *C*.

Angle *ABC* is the angle inside the triangle at *B*.

Angle *ABC* = 180° − (50° + 70°) = 60°

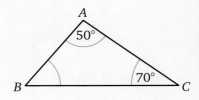

Example: Work out the size of angle *y*.

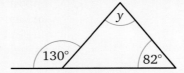

Solution: $y + 82° = 130°$

So $y = 130° − 82° = 48°$

Practise...   2.5 Angles and triangles   G F E D C

G

1   From the triangles below write down the letters of all triangles that are:

   a   equilateral   **c**   scalene

   b   isosceles   **d**   right-angled.

   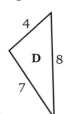

Not drawn accurately

E

2   Work out the size of the angles marked by letters.

   a   **c** 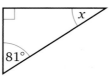   **e**

Not drawn accurately

   b   **d** 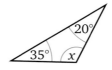   **f**

3   An isosceles triangle has a base angle of 56°.

Work out the size of the two other angles in the triangle.

4 Work out the angles marked by letters.

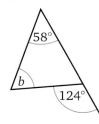
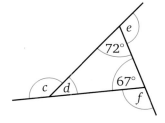

Not drawn accurately

5 Work out the angles marked with letters. Give a reason for each answer.

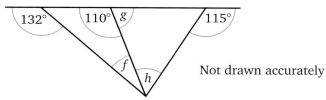

Not drawn accurately

6 Calculate the angles marked with letters. Give a reason for each answer.

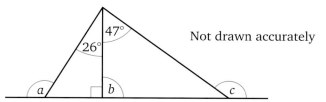

Not drawn accurately

7 The diagram shows a side view of a roof truss.

Work out the angles the beams form with the floor (angles x and y).

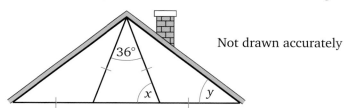

Not drawn accurately

8 Ben is working on a ladder as shown.

The angle between the ladder and the wall is 35°.

For Ben to use the ladder safely, the angle between the ladder and the ground must be between 70° and 80°.

Is Ben using the ladder safely?

Show working to justify your answer.

Not drawn accurately

2 Assess k!

1 What fraction of a full turn is:
   a 90°   b 135°   c 60°?

2 Here is a list of angles.

122°   17°   39°   242°   97°   305°   196°   82°

From this list, write down:

   a two acute angles   b two obtuse angles   c two reflex angles.

F

3   Work out the size of each of the marked angles.

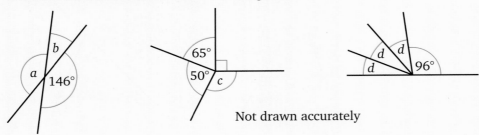

Not drawn accurately

4   Is *ABC* a straight line?

Show how you would decide

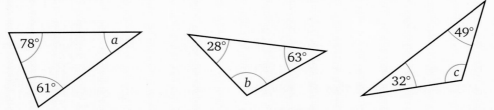

5   Work out the angle marked by each letter and state what type of angle it is.

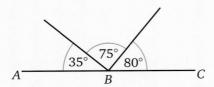

Not drawn accurately

6   Three angles form a straight line. The angles are *x*, 2*x* and 60°.

Work out the value of *x*.

7   Show that *ABC* is an isosceles triangle.

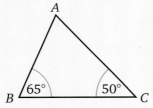

8   Triangle *ABC* has a right angle at *A* and angle *B* is 12° more than angle *C*.

Work out the sizes of angles *B* and *C*.

9   Work out the size of each of the marked angles.

Give a reason for each answer.

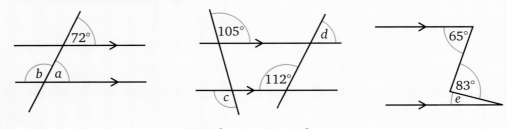

Not drawn accurately

10 For each diagram write down the three-figure bearing of P from Q.

a
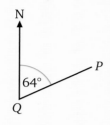

b
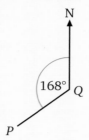

c
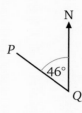

Not drawn accurately

11 The diagram shows the positions of three buoys A, B and C.

B is due east of A and C is due north of A.

The bearing of B from C is 129°.

Work out angle x.

What is the bearing of C from B?

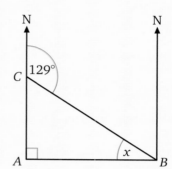

Not drawn accurately

12 Show that the line XA is parallel to BC.

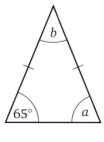

Not drawn accurately

⚠13 For each of the diagrams in Question 10, write down the bearing of Q from P.

AQA Examination-style questions 🖍

1 Two sides of this triangle are equal in length.

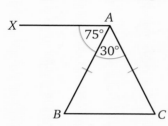

Not drawn accurately

a   What special name is given to this type of triangle? *(1 mark)*

b   i   Write down the value of a. *(1 mark)*
     ii   Work out the value of b. *(2 marks)*

AQA 2008

2 a   The diagram shows three angles on a straight line.

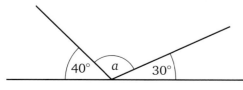

Not drawn accurately

Work out the value of a. *(1 mark)*

b   The diagram shows two intersecting straight lines.

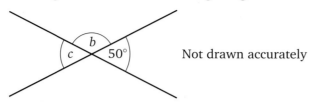

Not drawn accurately

    i   Work out the value of *b*.   *(1 mark)*
    ii Work out the value of *c*.   *(1 mark)*

c   The diagram shows a right-angled triangle.

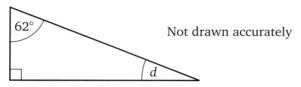

Not drawn accurately

Work out the value of *d*.   *(2 marks)*

AQA 2008

3   In the diagram *AB* is parallel to *CD*.

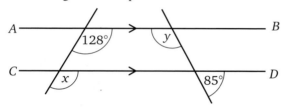

Not drawn accurately

  a   Write down the value of *x*. Give a reason for your answer.   *(2 marks)*
  b   Work out the value of *y*.   *(2 marks)*

AQA 2008

4   Rebecca has three rectangular sheets of paper.
She cuts each sheet into two pieces.
She now has six pieces, *A* to *F*, shown below.

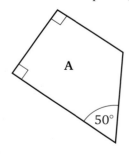
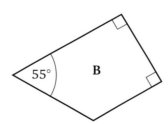
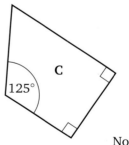

Not drawn
accurately

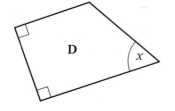
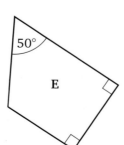

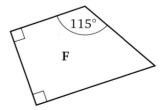

  a   Which piece is part of the same rectangle as *A*?   *(1 mark)*
  b   Which piece is part of the same rectangle as *B*?   *(1 mark)*
  c   Calculate the size of angle *x* on piece *D*.   *(2 marks)*

AQA 2006

Objectives

Examiners would normally expect students who get these grades to be able to:

F

simplify an expression such as $3x + 2x - x$

work out the value of an expression such as $4x - 3y$ when $x = 1$ and $y = 2$

E

simplify an expression such as $3x + 2 - 5x + 4$

understand the rules of arithmetic as applied to algebra, such as $x - y$ is not equal to $y - x$

work out the value of an expression such as $5x - 3y$ when $x = -2$ and $y = -3$

D

expand brackets such as $x(x + 2)$ in context

factorise an expression such as $x^2 + 4x$

C

expand and simplify an expression such as $x(2x + 1) - x(2x - 3)$.

Did you know?

Supermarket queuing theory

Supermarkets use queuing theory to decide how many people they need to work at the checkout at any time.

There are mathematical formulae used in queuing theory.

This is Little's theorem: $N = \lambda T$

Little's theorem is used in queuing theory.

    N stands for the average number of customers.

    λ is the average customer arrival rate.

    T is the average service time per customer.

Try testing out Little's theorem with real values to see if it makes sense.

Key terms

expression
term
like terms
simplify
unlike terms
substitution
expand
multiply out
factorise

You should already know:

✔ number operations and BIDMAS

✔ how to add and subtract negative numbers

✔ how to multiply and divide negative numbers

✔ how to find common factors

✔ about angle properties

✔ how to find the perimeter and area of shapes.

Learn... 3.1 Collecting like terms

An **expression** is a collection of **terms** and does not have a solution, e.g.

$3x - 2 + 2y - 4x + 1 + 5y$

There is more than one way of collecting **like terms** in an expression.

One way is to separate each term with a line. This helps keep the + and − signs in the right place.

$3x | - 2 | + 2y | - 4x | + 1 | + 5y$

Another way is to put circles around each term.

$3x - 2 + 2y - 4x + 1 + 5y$

Another way is to write like terms in columns and add them up.

$3x - 2 + 2y - 4x + 1 + 5y$

$$
\begin{array}{rrr}
3x & -2 & +2y \\
-4x & +1 & +5y \\
\hline
-x & -1 & +7y \\
\end{array}
$$

Another way is to underline each like term in a different colour.

$3x - 2 + 2y - 4x + 1 + 5y$

When you **simplify** an expression, you collect like terms.

A common mistake is to think that an x^2 term is the same sort of term as an x term. x^2 and x are **unlike terms**. The power makes a difference.

Sometimes you may want to rewrite the expression putting the like terms next to each other first.

$4x^2 + 3x - 3x^2$

$4x^2 - 3x^2 + 3x$

$x^2 + 3x$

Example: Find an expression for the perimeter of this triangle.

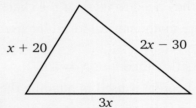

Solution: The sides add up to $x + 20 + 2x - 30 + 3x = 6x - 10$

AQA Examiner's tip

Use your own method for collecting terms that ensures you don't make mistakes with negative values. Remember, the sign is part of the term.

Practise... 3.1 Collecting like terms

G F E D C

F

1 Collect the like terms in each expression.

a $5p + 3p - p$

b $2a - 3a + 5a$

c $9c + 4c - c$

d $2x - 10x + 3x$

e $q - 11q + 8q$

f $20b - 7b - 13b$

g $6d - 3d + d - 5d$

h $10f - 4f - 3f - 2f$

F

2 Write down an expression for the perimeter of each shape.

a
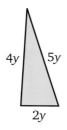

c
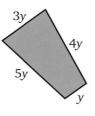

e
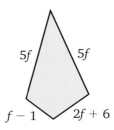

b
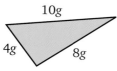

d
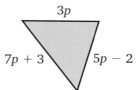

f
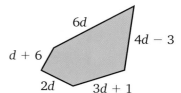

> **Hint**
> The perimeter of a shape is the total of the lengths of all the sides.

3 Write down an expression for the angles on each straight line.

a
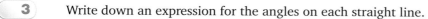

b

c

4 Write down an expression for the angles around the point in each diagram.

a
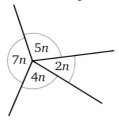

b
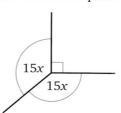

c
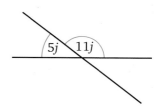

5 Simplify:

a $3d + 8d - 4d$

b $2x - 3x + 5x$

c $-3w + 10y - 5w + 3$

d $4c - 3 + 8d + c$

e $g + 2f - 3g - 4$

f $4fg - 3g + 2 - 5fg$

g $3b + 2a - 3b + a$

h $12x - 15y - 5 + 10x$

i $17 - 10m + 5 - 3n + 12$

6

a   Show that $5x^2 - 4x^2 + x^2 = 2x^2$

b   Show that $a^2 - a + 3a^2 - 5a = 4a^2 - 6a$

c   i   Find the mistake Shannon made when she collected the like terms in this expression:
$2f - 3f^2 - 2f^2 + 5f = 2f$

ii   Give the correct answer to the simplification.

7 Simplify:

a $2x^2 - 3x^2 + x^2$

b $-5y^2 + y^2 - 2y - y$

c $2h - 5 + h^2 - h$

d $10v + 10v^2 - 15v^2 + v$

e $-3p^2 + 2p^2 + 4p^2 + 4t^2$

f $5t^2 + m^2 - 3m^2 - 6t^2$

8   Nilima collects one term from each column and simplifies her answer.
Her route is shown in the diagram.

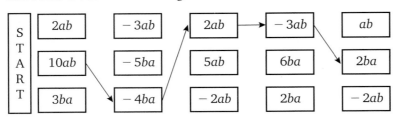

Hint

Remember: ab and ba are equivalent.

Nilima's route and simplified answer are: $10ab - 4ba + 2ab - 3ab + 2ba = 7ab$

a   Choose your own route and simplify your expression in the same way as Nilima.

b   Find a different route that gives the same answer as Nilima.

c   Find a route that simplifies to give the answer ab.

Learn...   3.2 Substitution

When each of the letters in an expression represents a given number, you can find the value of that expression. This is called **substitution**.

If you are substituting negative numbers be careful to follow the rules for the addition, subtraction, multiplication and division of positive and negative numbers. The same rules apply in arithmetic and in algebra.

The rules of arithmetic and the rules of algebra

$x + y$ is the same as $y + x$ $x - y$ is **not** the same as $y - x$
$x + -y$ is the same as $x - y$ $x - -y$ is the same as $x + y$

$a \times b$ is the same as $b \times a$ $\dfrac{a}{b}$ is **not** the same as $\dfrac{b}{a}$

$a \times -b = -ab$ $\dfrac{-a}{b} = -\dfrac{a}{b}$

$-a \times -b = ab$ $\dfrac{-a}{-b} = \dfrac{a}{b}$

Example: Substitute the given values into the expression.
Simplify your answer.

     a $p + q$ $p = 7$ $q = 2$ **b** $cd + 3c$ $c = 3$ $d = -4$

Solution: **a** $p + q = 7 + 2 = 9$ **b** $cd + 3c = 3 \times -4 + 3 \times 3$
                                                    $= -12 + 9 = -3$

Practise...   3.2 Substitution    G F E D C

1   Substitute the values into each expression. Simplify your answer.

     a $x + y$ $x = 2$ $y = 5$

     b $a - b$ $a = 4$ $b = 7$

     c $2p + q$ $p = 0.5$ $q = 1$

     d $3m - 2t$ $m = 3$ $t = 4$

     e ab $a = 4$ $b = 15$

     f $3cd$ $c = 2$ $d = 7$

     g $5gh + g$ $g = 2$ $h = 4$

     h $10 - cd$ $c = 2$ $d = 0.5$

     i $\dfrac{x}{2} + 2y$ $x = 10$ $y = 1$

     j $ab + \dfrac{b}{2}$ $a = 3$ $b = 4$

F

2 Find the value of each expression, using the values given for x and y.

a $30° + 2x - 3y$ $x = 15°$ $y = 5°$

b $120° - 3y - x$ $x = 10°$ $y = 15°$

c $4x - 10° - 5y$ $x = 45°$ $y = 7°$

d $5y - 6x + 112°$ $x = 12°$ $y = 32°$

e $\dfrac{360°}{2x}$ $x = 5°$

f $\dfrac{180°}{(2x - y)}$ $x = 20°$ $y = 10°$

3 Chris has completed this pyramid puzzle by adding the expressions in the blocks next to each other and writing the result in the block above.

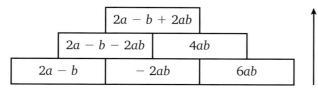

Use $a = 3$ and $b = 4$ to find the value of each block in the pyramid.

E

4 Tracey and Debbie are practising substitution. They have different expressions and values to substitute but their answers should be the same.

These are the first expressions they have worked out.

Tracey $2k + 4m$ $k = 5, m = -2$
           $2(5) + 4(-2)$
           $10 - 8$
           2

> **Hint**
> Put brackets in to help avoid mistakes with negative numbers.

Tracey and Debbie both get the answer 2.

Debbie $3k - 5m$ $k = 4, m = 2$
           $3(4) - 5(2)$
           $12 - 10$
           2

Now find the answers to Tracey and Debbie's questions to find out which give the same answers and which don't.

	Tracey			Debbie	
a	$3p - 2q$	$p = -2, q = 1$		$4p + q$	$p = 0.5, q = 1$
b	$4x - \dfrac{3y}{2}$	$x = 2, y = -1$		$7x - 2y$	$x = 0.5, y = -0.5$
c	cd	$c = -\frac{1}{2}, d = 4$		$\dfrac{4}{cd}$	$c = -1, d = 2$
d	$\dfrac{4a}{b}$	$a = 5, b = -2$		$2a - b$	$a = -6, b = -2$

5 Ali has been carrying out an investigation into area.

In this rectangle the length added to the width is 10 cm.

> **Hint**
> The area of a rectangle is width × length.

a Using whole numbers for the width and length, find the greatest area. Make sure you write down all the numbers you try.

b Ali wants to try negative numbers in his investigation. Why won't that work?

E

6   **a**   **i**   Write down and simplify an expression for the perimeter of this pentagon.

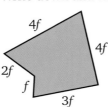

4f

4f

2f

f

3f

ii   Find the perimeter of the pentagon when $f = 2$ cm.

b   The length of each side of this pentagon is increased by 3 cm.

   i   Write an expression for the length of each side of the new pentagon.

   ii   Write down and simplify an expression for the perimeter of the new pentagon.

   iii   Find the perimeter if $f = 2$ cm

D

7   Write an expression for the perimeter of each polygon.   Not drawn accurately

a

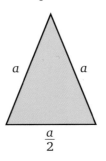

a a

$\dfrac{a}{2}$

d

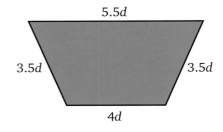

5.5d

3.5d   3.5d

4d

b

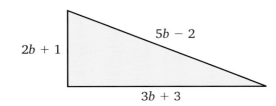

$5b - 2$

$2b + 1$

$3b + 3$

e

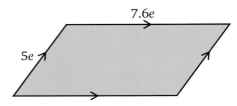

7.6e

5e

c

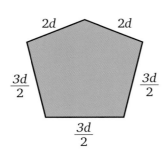

2d   2d

$\dfrac{3d}{2}$ $\dfrac{3d}{2}$

$\dfrac{3d}{2}$

f

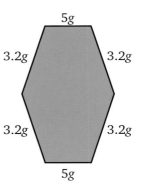

5g

3.2g   3.2g

3.2g   3.2g

5g

8   These two rectangles have the same area.
Both x and y are whole numbers.

A

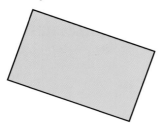

B

Not drawn accurately

width $= x + 1$
length $= x - 1$
If $x = 5$, find the value of y.

length $= 2y$
width $= y - 1$

9   In Kylie's pyramid puzzle, the expressions in blocks next to each other are added together to give the expression in the block above.

a   Copy and complete the expressions in Kylie's pyramid puzzle.

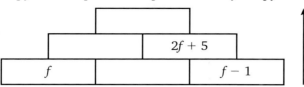

b   Work out the value of each block in Kylie's pyramid if $f = 1.5$

c   Design your own pyramid puzzle and try it out on your friends.

10   Here are three expression cards.

a   Which card gives the greatest answer when:

   i $a = 2$ and $b = 3$?

   ii $a = 2$ and $b = -3$?

$a - b$	$\dfrac{a}{b}$	$3a$
card 1	card 2	card 3

b   Can you find a pair of values for which cards 2 and 3 are equal?

11   In each of the following expressions the value of n is 5 and the value of m is -2.

Copy each pair of expressions and between them write $>$, $<$ or $=$ to make the statement correct.

a $m - n$ ..... $2m$ c $5m + 7$ ..... $10 - n$

b m^2 ..... $3n - 10$ d $n^2 - 10$ ..... $m^2 + 10$

Learn...   3.3 Expanding brackets and collecting like terms

When you **expand** brackets, **all** the terms inside the brackets must be multiplied by the term outside the brackets.

You will be given the instruction **expand** or **multiply out**.

If there is more than one bracket in the expression, each is done separately before collecting terms.

If there are terms not included in the bracket they are not included in the expansion.
They are collected after the brackets have been expanded.

Example: Expand $3(x - 2)$.

Solution: One method for expanding brackets is the grid method.

You have seen other methods in Unit 2.

×	x	$- 2$
3	$3x$	$- 6$

$x - 2$ goes here. The terms are put into separate boxes.

3 goes here

$3(x - 2) = 3x - 6$

One way of checking your answer is to substitute a value for x into the expression before the expansion and the expression after the expansion. If you get the same answer then your expansion is correct.

e.g.   If $x = 5$
then $3(5 - 2) = 3 \times 3$ ⟵ Before the expansion
                    $= 9$

and $3 \times 5 - 6 = 15 - 6$ ⟵ After the expansion
                    $= 9$

Example: Multiply out $5a(2a + 1)$.

You can use other methods for expanding brackets.

Solution: Using the grid method:

×	2a	+ 1
5a	10a²	+ 5a

Remember that $a \times a = a^2$

$5a(2a + 1) = 10a^2 + 5a$

You can check your expansion by substituting values for x.

Example: Expand and simplify $9x - 2(x - 4)$.

Solution: Note that $9x$ is not included in the bracket.

The part of the expression that needs to be expanded is $- 2(x - 4)$.

Only use the grid to expand the brackets. $9x$ is not used at this point.

×	x	− 4
−2	−2x	+ 8

Remember the sign stays with the number.

$-2(x - 4) = -2x + 8$

Put the $9x$ back into the expression and then simplify.

$9x - 2(x - 4) = 9x - 2x + 8$
$= 7x + 8$

Example: Expand and simplify $4(3y + 2) - 5(y - 3)$.

Solution: Separate the expression into two brackets to carry out the expansion.

Expand $4(3y + 2)$

×	3y	+ 2
4	12y	+ 8

$4(3y + 2) = 12y + 8$

Expand $- 5(y - 3)$

Remember the minus sign is included here

×	y	− 3
−5	−5y	+ 15

$-3 \times -5 = +15$

$-5(y - 3) = -5y + 15$

Step 3:   Put the two answers together and collect like terms.

$4(3y + 2) - 5(y - 3) = 12y + 8 - 5y + 15$
$= 7y + 23$

Example: Write an expression for the area of this rectangle. Expand your answer.

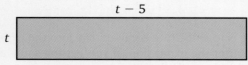

$t - 5$

t

Solution: Area of a rectangle is width × length.

Area $= t(t - 5)$

Put the length in a bracket so both terms are multiplied by t.

Expand the brackets using your own method.

Area $= t^2 - 5t$

There are no terms to simplify as t^2 and t are different types of term.

AQA *Examiner's tip*

You can use any successful method to expand brackets.

Practise... 3.3 Expanding brackets and collecting like terms

G F E D C

D

1 Expand:

a $3(x + 4)$ c $8(2 - c)$ e $5(5d - 1)$ g $3(10v + 7)$

b $5(y - 2)$ d $3(2p + 5)$ f $7(2 - 2f)$ h $11(7 + 3m)$

2 Write an expression for the area of each of these rectangles.
Expand your answer.

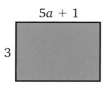

a $5a + 1$, 3

d $1.5d$, $d - 2$

b $b - 5$, 2

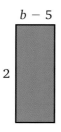

e $2.5e$, $10 + e$

c $4c + 3$, 6

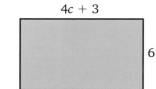

f $3 - f$, $5.5f$

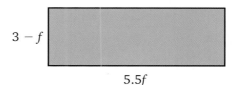

Hint

Remember to use brackets and multiply out all the
terms. Then collect like terms and simplify them.

3 Dora and Jim are writing number puzzles using symbols.
They use n to represent the missing number.

Dora says 'Think of a number, add two and then multiply the answer by 5.'

Jim writes his answer as $n + 2 \times 5$.

a Write down the mistake Jim has made and re-write his answer correctly.

b Write each of these number puzzles as an expression, using n for the
missing number.

  i 'Think of a number, subtract 8 and then multiply the answer by 5.'

  ii 'Begin with 10 and subtract the number, multiply the answer by 3
  and then add 3 times the number.'

c Travis and his friends wrote these expressions for number puzzles. For
each one, expand the brackets and collect like terms.

  i $5(n + 2) - 10$

  ii $3n - 2(n - 1)$

  iii $5(n + 1) + 2(3 - n)$

  iv $7(3 - 2n) + 10n$

  v $3(2 + 3n) - 2(3 + 4n)$

  vi $2(n + 1) + 5(n - 2) - 4n$

C

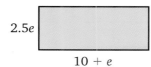

Bump up your grade

At Grade C you may be asked to
expand more than one bracket
before simplifying.

C

4 Find an expression for the shaded area of each shape. Expand and simplify your answer.

Hint Find one area then add or subtract the other.

a
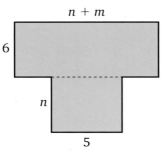

d
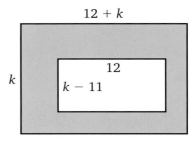

b
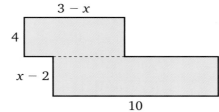

e
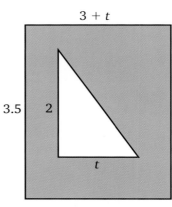

Hint Remember the area of a triangle $= \frac{1}{2}$ base × height

c
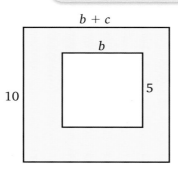

f
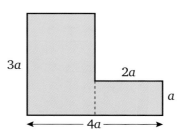

5 Find the missing numbers to make each pair of expressions equal.

a $2(a + 4) + 3(a - 1)$ $?(a + 1)$

b $5(1 - b) - 3(b - 3)$ $?(7 - 4b)$

c $2(x + 1) - 5(x + ?)$ $-3(x + 6)$

d $?(2x - 1) + 2(x - 3)$ $10(x - 1)$

e $5(1 - x) + ?(4x + 5)$ $3(x + 5)$

6 In each diagram the value of the perimeter and the numerical value of the area are the same. In each question part, which value of x gives the same answer for the perimeter and the area?

a
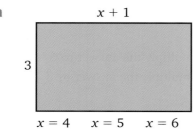
$x = 4$ $x = 5$ $x = 6$

b
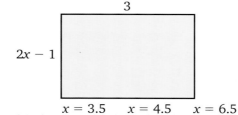
$x = 3.5$ $x = 4.5$ $x = 6.5$

c
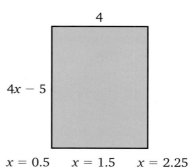
$x = 0.5$ $x = 1.5$ $x = 2.25$

Hint Start by writing an expression for the perimeter and an expression for the area.

7   Here are two rectangles.

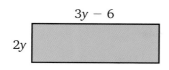

$3y - 6$

$2y$

$6y$

$y - 2$

Show that the two rectangles have equal area.

8   Row A $\boxed{3}$ $\boxed{4x}$ $\boxed{2}$ $\boxed{3x}$ $\boxed{5x}$ $\boxed{-5}$ $\boxed{2x}$ $\boxed{-x}$

Row B $\boxed{3 - 2x}$ $\boxed{x - 2}$ $\boxed{2x + 1}$ $\boxed{3 - x}$ $\boxed{x - 1}$ $\boxed{x + 2}$ $\boxed{2 - x}$ $\boxed{3x - 1}$

Jess chooses two terms from row A and two expressions from row B to make this expression:

$3(3 - x) + 3x(x - 2)$

Jess simplifies the expression in two steps:

$9 - 3x + 3x^2 - 6x$

$3x^2 - 9x + 9$

a   Choose your own four terms and expressions to make an expression and simplify it in the same way Jess has.

b   The following shows Jess's working when simplifying three other expressions. Find the missing terms from the cards above, then copy and complete her working.

   i $?(2x + 1) - 5(x + 2)$ **ii** $4x(?) - x(2 - x)$ **iii** $5x(x - 1) + ?(?)$

     $= 4x^2 + 2x - 5x - 10$ $= ? - ? - 2x + x^2$ $= 5x^2 - ? + 6 - ?$

     $=$ $= 13x^2 - 6x$ $= 5x^2 - ? + 6$

Learn...   **3.4 Factorising expressions**

Factorising is the opposite of expanding.

You will usually be given the instruction **factorise**.

Example: An expression for the area of this rectangle is $3xy + 6y$.

Factorise the expression to find possible dimensions of the rectangle.

Solution: You can use the grid method **in reverse** to factorise this expression.

Step 1: Find the common factor of the two terms in the expression.

$3y$ is the common factor because 3 is a factor of 3 and 6, and y is a factor of xy and y.

3y	3xy	+ 6y

Put $3y$ in the grid along with the two terms in the expression.

Link

You can find more on factorising in Unit 2, Chapter 5.

Step 2: Divide each term by the common factor.

$3xy \div 3y = x$

$6y \div 3y = 2$

	x	+ 2
3y	3xy	+ 6y

Put x and $+ 2$ into the grid.

Step 3: Read the values from the grid to give the answer.

$3xy + 6y = 3y(x + 2)$

Practise... 3.4 Factorising expressions

D

1   Factorise each expression.

a $8c + 4$ e $20x + x^2$ i $4n + 18n^2$

b $12d - 15$ f $y^2 - 5y$ j $15kl + 27k^2$

c $20 - 10p$ g $12xy - 9y^2$ k $13f^2 - 65fg$

d $24 + 18k$ h $b^2 + 9ab$ l $36j^2k - 30jk^2$

2   In each question, one side of the rectangle and the expression for the area has been given.

Use factorising to find an expression for the length of the other side.

a

area = $6x - 21$ 3

b $5x$

area = $20x^2 - 25x$

c

area = $7p^2 - 5p$ p

> **AQA Examiner's tip**
>
> Always factorise fully. The highest common factor of all the terms in an expression should be outside the bracket.

d $4t + 5$

area = $100t^2 + 125t$

e k

area = $2kq - 3kr$

3   These expressions have been factorised.

Find the missing numbers or terms in each question part.

a $10p - 8 = ?(5p - 4)$ d $20pq - 15q = 5q(\ ?\ )$

b $12x - 15y = ?(4x - 5y)$ e $24x^2 + 9x = ?x(\ ?\ )$

c $11ab + 7bc = ?(? + 7c)$ f $19fh + 38fgh = 19fh\ (? + ?)$

4   Factorise fully the expressions for the area of these rectangles.

Use your answers to give possible dimensions of each rectangle.

a

area = $5g + 10gt^2$

D

b   area $= 3x^2 + 21xy^2$

c

area $= 6p^2q^2 - 2pq^3$

Hint
$p^3 \div p^2 = p$

d   area $= 5a^2bc^2 + 15ab^2c$

5   Chris and Sue both factorise this expression but get different answers.

$12xy^2 - 18x^2y$

Chris's answer is $2x(6y^2 - 9xy)$

Sue's answer is $6xy(2y - 3x)$

Who is correct? Give reasons for your answer.

6   These expressions have been factorised fully. Some are wrong and some are right.

For each question part, say whether the answer is wrong or right.

If it is wrong, give the correct answer.

a $3ap - 9p = 3p(a - 3)$

b $12f^2 - 18f = 3(4f^2 - 6f)$

c $36 - 4t^2 + 12t = 4(9 - t^2 + 3t)$

d $15x^2y^2 - 20x^2y = 5x(3xy^2 - 4xy)$

e $55k - 44klm^2 = 11k(5 - 4lm^2)$

7   In this factor puzzle the first expression is factorised. The factors are added.
This is repeated until the expression cannot be factorised again.

C

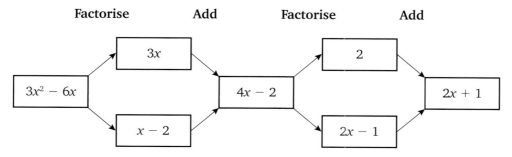

a   Copy and complete this factor puzzle.

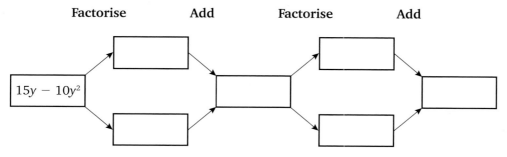

b   Complete a factor puzzle for each of these starting expressions.

i $7a^2 - 14a$ **ii** $4p^2 - 12p$ **iii** $28c - 35c^2$ **iv** $20q - 25q^2$

3   Assess (k!)

1   Write down and simplify an expression for the perimeter of this triangle.

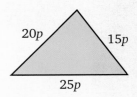

2   **a**   Write down and simplify an expression for the perimeter of this pentagon.

     b   Work out the perimeter when:

        i $p = 2$ cm   **ii** $p = 10$ cm   **iii** $p = 0.5$ cm.

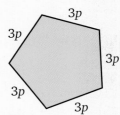

3   Work out the value of $2h - 3$ when:

     a $h = 2$ **b** $h = 10$ **c** $h = -2$ **d** $h = -4$

4   Simplify:

     a $3a - 5 + 4a + 2$ **b** $10ab - 3ab + 4ab - 2ab + ab$ **c** $3d + 2d^2 - 3 - 4d$

5   Write an expression in terms of b for the area of this rectangle. Write your expression in its expanded form.

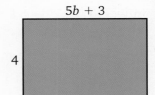

6   Expand and simplify:

     a $12 + 4(x - 2)$ **b** $3(5 - 3p) - 2p$

7   Factorise fully:

     a $14f + 21$ **b** $24x - 18xy$

8   Factorise fully:

     a $5x + x^2$ **b** $y^2 - 3y$ **c** $7pq^2 - 56pq$

9   Expand and simplify $2(3x + 1) - 3(4 - x)$.

10   Write an expression for the yellow area of this shape.

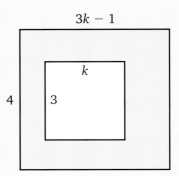

AQA Examination-style questions (k!)

1   **a**   Multiply out $a(b + c)$.   *(1 mark)*

     b   Work out the value of $xy + xz$ when $x = 27$, $y = 3$ and $z = 7$ *(3 marks)*

AQA 2008

4 Percentages and ratios

Objectives

Examiners would normally expect students who get these grades to be able to:

F

understand that percentage means 'number of parts per 100' and use this to compare proportions

E

work out a percentage of a given quantity

D

increase or decrease by a given percentage

express one quantity as a percentage of another

use ratio notation, including reduction to its simplest form and its links to fraction notation

solve simple ratio and proportion problems, such as finding and simplifying a ratio

C

work out a percentage increase or decrease

solve more complex ratio and proportion problems

solve ratio and proportion problems using the unitary method.

Key terms

percentage	ratio
unitary method	unitary ratio
VAT (Value Added Tax)	proportion
rate	

Did you know?

The origins of mathematical symbols

15th century	17th century	Now
P⌐°	° / °	%

Did you know that the symbol % for percent and : for ratio were not introduced until relatively recently? The % sign probably developed from the symbol shown above. This appeared in an Italian manuscript dating from 1425. In Italian 'each hundred' is 'per cento', in French it is 'pour cent' and in English 'per cent'.

An Englishman, William Oughtred, was the first to use a colon : for ratios in his book, *Canones Sinuum* in 1657.

Find out when other symbols such as $+$, $-$, $\times$, $\div$ and $=$ were introduced. What did they use before then?

You should already know:

- ✔ how to add, subtract, multiply and divide simple numbers with and without a calculator
- ✔ about place values in decimals (for example, $0.7 = \frac{7}{10}$, $0.07 = \frac{7}{100}$)
- ✔ how to put decimals in order of size
- ✔ how to simplify fractions by hand and using a calculator
- ✔ how to write a fraction as a decimal and vice versa
- ✔ how to change a percentage to a fraction or decimal, and vice versa.

Learn... 4.1 Finding a percentage of a quantity

Sometimes writing a **percentage** as a fraction can help you find a percentage of a quantity.

For example, to shade 75% of this shape:

$75\% = \dfrac{75}{100} = \dfrac{3}{4}$, so shade three quarters of the shape

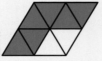

You can use your calculator's fraction key to simplify fractions.

(As $\frac{3}{4} = \frac{6}{8}$, any 6 of the 8 triangles could be shaded.)

To write a percentage as a fraction or decimal, divide by 100:

$75\% = 75$ out of $100 = \dfrac{75}{100}$

Also $75\% = 75 \div 100 = 0.75$

You can use your calculator to work out percentages of other quantities.

This is the decimal equivalent of the percentage you wish to find.

To find a percentage of a quantity on a calculator:
● **divide** the quantity **by 100** (to find 1%)
● then **multiply by the percentage** you need.

Finding 1% first is sometimes called the **unitary method**.

Or use a **multiplier:**
● write the percentage as a multiplier
● then **multiply** by the quantity.

To increase or decrease by a given percentage:
● find the percentage of the quantity (as above)
● for a percentage increase, add to the original quantity
   for a percentage decrease, subtract from the original quantity

Or use a **multiplier:**
● write the new quantity as a percentage of the original quantity
● convert this percentage to a multiplier
● **multiply by the original quantity**.

Multipliers

You can use a multiplier to find a percentage of a quantity or to increase or decrease a quantity by a percentage.

For example:
● **to find 35% of a quantity**, the multiplier is $35 \div 100 = $ **0.35**
● **to increase a quantity by 35%**, the multiplier is $135 \div 100 = $ **1.35** ◄——— because $100\% + 35\% = 135\%$
● **to decrease a quantity by 35%**, the multiplier is $65 \div 100 = $ **0.65** ◄——— because $100\% - 35\% = 65\%$

Example: Find $26\frac{1}{2}\%$ of £9.47.

To find $26\frac{1}{2}\%$
Use 26.5 if you prefer.

Solution: $26\frac{1}{2}\%$ of £9.47 $= £9.47 \div 100 \times 26\frac{1}{2}$
$\qquad = £2.50955$
$\qquad = £2.51$ To find 1%
$\qquad$ (to the nearest penny)

Round to the nearest penny.

Alternative method using multiplier
$26\frac{1}{2}\% = 26.5\% = 0.265$ as a decimal
$26\frac{1}{2}\%$ of £9.47 $= 0.265 \times 9.47$
$\qquad\qquad = £2.51$
$\qquad$ (to the nearest penny)

Example: Increase £72 by 30%

Solution: 30% of £72 $= 72 \div 100 \times 30$
$\qquad = 21.6$

To find 1%, then 30%

New amount $= 72 + 21.6$
$\qquad\qquad = £93.60$

Alternative method using multiplier
New amount $= 100\% + 30\% = 130\%$
Multiplier $= 1.30$ or 1.3

New amount $= 1.3 \times 72$
$\qquad\qquad = £93.60$

AQA *Examiner's tip*

Take care to write money correctly.
Here the answer is £93.60, not £93.6

Example: Decrease 3.25 litres by 42%

Solution: 42% of $3.25 = 3.25 \div 100 \times 42$
$\qquad\qquad = 1.365$

New amount $= 3.25 - 1.365$
$\qquad\qquad = 1.885$ litres

Alternative method using multiplier
New amount $= 100\% - 42\% = 58\%$
Multiplier $\quad = 0.58$

New amount $= 0.58 \times 3.25$
$\qquad\qquad = 1.885$ litres

Does your calculator have an 'Ans' key?
You can use it instead of entering 1.365 again here.

4.1 Finding a percentage of a quantity

1 Copy each shape onto squared paper. Shade the given percentage.

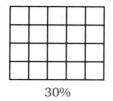

a   b   c

75%   30%   90%

2   **a**   Use one of the methods shown in Learn 4.1 to work out:

i	20% of £380	iii	62% of 3500 litres	v	37% of 600 g
ii	25% of £7000	iv	85% of 40 km	vi	18% of 750 ml.

b   Use a different method to check your answers.

3   **a**   Write down the multiplier for working out 40% of a quantity.

> **Hint**
>
> This is how you find the multiplier for working out 40% of a quantity:
>
> Multiplier = 40 ÷ 100 = 0.4

b   Copy and complete this table of multipliers.

To find	40%	1%	15%	24%	9%	6%	12.5%	3.5%	225%
Multiply by		0.01		0.24		0.06			

c   Use the multipliers from your table to work out:

i	40% of 250	iv	24% of £900	vii	12.5% of £140
ii	1% of 420	v	9% of £50	viii	3.5% of £72.60
iii	15% of 640	vi	6% of £230	ix	225% of £42.70.

4   A school has 60 teachers.
85% of the teachers work full time; the rest work part time.

a   How many teachers work full-time?

b   How many teachers work part-time?

5   Louise says '60% of £19 is eleven pounds and four pence.'
Is she correct?
Show working to support your answer.

6   **a**   Write down the multiplier for increasing a quantity by 10%

> **Hint**
>
> After an increase of 10%, the new quantity is 100% + 10% = 110% of the original quantity.
>
> The multiplier = 110 ÷ 100 = 1.10

b   Copy and complete these tables.

To increase by	10%	20%	5%	7.5%
Multiply by			1.05	

To decrease by	10%	20%	5%	7.5%
Multiply by		0.8		

c   Use the multipliers from your table to work out:

i	90 increased by 10%	v	£375 decreased by 10%
ii	150 g increased by 20%	vi	£549 decreased by 20%
iii	240 m increased by 5%	vii	620 kg decreased by 5%
iv	£364 increased by 7.5%	viii	480 litres decreased by 7.5%

D

7
   a   Increase 250 m by 40%   **d**   Decrease 37.5 litres by 12%

   b   Increase 80 kg by 70%   **e**   Increase £54.60 by 43%

   c   Decrease 24 miles by 5%   **f**   Decrease £180 by 62.5%

Check your answers using a different method.

8
   The price of a magazine is £2.50.

What is the new price after a 6% increase?

Check your answer using a different method.

> **AQA** *Examiner's tip*
>
> Always **check** your answers. You can do this by using a different method or by checking to see that the answer is reasonable.

9
   The table gives the original prices of some sports equipment. The shop reduces these prices by 20% in a sale.

Find the new prices.

Item	Original price
Football	£15.90
Tennis racket	£65.75

10
   The population of a city is 275 300.
It is expected to increase by 3% next year.

What is the expected population next year?

Give your answer to the nearest hundred.

11
   The prices of each of these items are given excluding **VAT**.
Find the cost of each item including VAT at the **rate** given.

a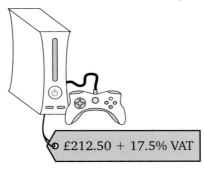

£212.50 + 17.5% VAT

c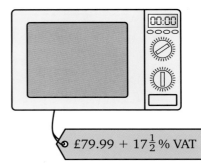

£79.99 + $17\frac{1}{2}$ % VAT

b

Cost of electricity:
£134.92 + 5% VAT

⚠12
   Three quarters of a million people sat GCSE Mathematics in the UK last year.
Of these people, 51% were female.

How many more females than males sat GCSE Mathematics in the UK last year?

⚙13
   A sponsored swim raised £480 for charity.

The organisers give 35% to a children's charity and 25% to an animal sanctuary.
They give the rest to a local hospice.

How much money do they give to the local hospice?

You **must** show your working.

⚙14
   A supermarket has offered Paul a job.

He can choose to work on Saturday or Sunday.
The table gives the hours of work.
He earns £6.40 per hour on Saturday. He earns 25% more per hour on a Sunday than he does on a Saturday.
On each day he has a 1 hour *unpaid* lunchbreak.

On which day does Paul earn more?

	Hours
Saturday	9am – 5pm
Sunday	10am – 4pm

15 A jar contains 80 sweets. The sweets are red, yellow, green, blue or orange.

25% of the sweets are red.
20% of the sweets are yellow.
There are 10% more orange sweets than red sweets.
50% of the remaining sweets are blue.

How many green sweets are there?

Learn... 4.2 Writing one quantity as a percentage of another

To write one quantity as a percentage of another:
- divide the first quantity by the second; this writes them as a decimal (or write the first quantity as a fraction of the second)
- then multiply by 100% to change the decimal or fraction to a percentage

To write an increase or decrease as a percentage:
- find the difference between the old quantity and the new quantity; this gives the increase or decrease
- divide the increase (or decrease) by the **original** amount or write the increase (or decrease) as a fraction of the **original** amount
- then multiply by 100% to change the decimal or fraction to a percentage

Sometimes you may need to write one part as a percentage of the whole amount (as in the first example below).

The quantities must always be in the **same units**.

You can also use this method to find a percentage profit or loss.

Example: Last season a school's football team won 18 matches, drew 4 matches and lost 10 matches.

  a What percentage of the matches were won?

  b What percentage were drawn?

  c What percentage were lost?

Solution: You must write each number as a percentage of the total number of matches.

The total number of matches = 18 + 4 + 10 = 32

  a percentage of games that were won = $\frac{18}{32} \times 100\% = 18 \div 32 \times 100\% = 56.25\%$

  or using the fraction key: percentage won = $\frac{18}{32} \times 100\% = 56\frac{1}{4}\%$

  b percentage of games that were drawn = $\frac{4}{32} \times 100\% = 4 \div 32 \times 100\% = 12.5\%$

  or using the fraction key: percentage drawn = $\frac{4}{32} \times 100\% = 12\frac{1}{2}\%$

  c percentage of games that were lost = $\frac{10}{32} \times 100\% = 10 \div 32 \times 100\% = 31.25\%$

  or using the fraction key: percentage lost = $\frac{10}{32} \times 100\% = 31\frac{1}{4}\%$

Add the percentages to check the answer: 56.25% + 12.5% + 31.25% = 100% ✓

Example: Before Heidi joined a gym, she weighed 72 kg. She now weighs 69 kg. What is the percentage decrease in her weight?

Solution: Decrease = 72 − 69 = 3 kg

Percentage decrease = $\frac{3}{72} \times 100\% = 3 \div 72 \times 100\% = 4.166...\%$

or using the fraction key $\frac{3}{72} \times 100 = 4\frac{1}{6}\%$

So the percentage decrease is 4.2% (to one decimal place).

AQA Examiner's tip

Remember to divide by the **original** amount.

Example: Jack's pay rate has gone up by 45 pence to £8.55 per hour.
Find the percentage increase.

Solution: The increase in Jack's hourly pay rate = 45 pence

To work in pence, use
£8.55 = 855 pence

His original hourly pay rate was 855 − 45 = 810 pence

Percentage increase = $\frac{45}{810} \times 100\% = 45 \div 810 \times 100\% = 5.555...\%$

The percentage increase in Jack's hourly pay rate = 5.6% (to 1 d.p.)

or $\frac{45}{810} \times 100\% = 5\frac{5}{9}\%$

You get the same answer if you work in pounds:
percentage increase = 0.45 ÷ 8.10 × 100% = 5.6% (to 1 d.p.)

AQA *Examiner's tip*

You must use the **same units** in your division sum.

4.2 Writing one quantity as a percentage of another

Practise...

G F E D C

When answers are not exact, round them to one decimal place.

D

1   Write the first quantity as a percentage of the second quantity.

  a   96p,   £3

  b   £22 500,   £90 000

  c   34 kg,   85 kg

  d   60 cm,   5 m

  e   270 g,   5 kg

  f   350 cm,   2 m

Hint

1 m = 100 cm
1 kg = 1000 g

2   In a class, 21 out of the 28 students walk to school.
What percentage of the class walk to school?

3   A farmers' market will be held on 8 days in September.
What percentage of the month is this?

4   A local authority collected 88 000 tonnes of waste for recycling last year.
9000 tonnes of this was glass.
What percentage of the waste was glass?

5   A school's athletic team is made up of 16 girls and 19 boys.
What percentage of the team are:

  a   girls

  b   boys?

C

6   A garden centre buys plants for 56 pence each.
It sells them for 99 pence each.
Work out the percentage profit.

Hint

The percentage profit is the percentage increase in price.

7 Sharon buys a motorbike for £3400.
She sells it a year later for £2900.
Work out her percentage loss.

Hint
The percentage loss is the percentage decrease in price.

8 This year a school has 1237 students. Last year there were 1329 students.
Find the percentage decrease in the number of students in the school.

Bump up your grade
To get a Grade C you must be able to write an increase or decrease as a percentage of the **original** amount.

9 The price of a packet of biscuits goes up from 98 pence to £1.09.
Find the percentage increase in the price.

10 A furniture shop reduces its prices in a sale.

a Work out the percentage reduction in the price of:
   i   the table
   ii one chair.

b Dan buys a table and four chairs.
Work out the percentage reduction in the total price.

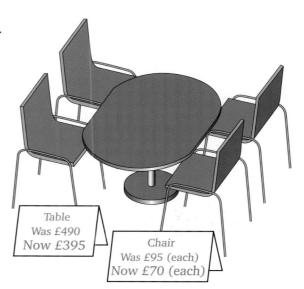

Table
Was £490
Now £395

Chair
Was £95 (each)
Now £70 (each)

11 The cost of Greg's car insurance has gone up from £420 to £480.
Greg works out $60 \div 480 \times 100$. He says the cost has increased by 12.5%

a What mistake has he made?

b What is the actual percentage increase in the cost of the insurance?

⚠12 Write:

a eighty thousand as a percentage of two million

b £75 million as a percentage of £5 billion

c 90 cm as a percentage of 1.75 m

d 750 g as a percentage of 3.6 kg

e 37.5 hours as a percentage of 1 week.

Hint
1 billion = 1000 million

⚠13 A manufacturer makes a rectangular rug that is 160 cm long and 120 cm wide.
The manufacturer decides to increase both dimensions of the rug by 25%

Find the percentage increase in:

a the perimeter of the rug

b the area of the rug.

14 The table gives the number of visits to some countries made by UK residents in 2004 and 2008.

Compare the percentage changes in the number of visits made to each of these countries.

AQA Examiner's tip

When you are asked to compare sets of data with a different amount of data in each set you should use percentages.

Visits abroad by UK residents (thousands)		
	2004	**2008**
Bulgaria	297	360
Greece	2709	2096
Italy	2974	3372
Slovakia	38	170
Turkey	1124	1936

Source: Travel Trends 2008 (Crown copyright)

15 A newspaper gives a table showing the changes in the quantities of milk and cream used between 2006 and 2010.

Work out the percentage changes and comment on your results.

Food	Average per person per week	
	2006	**2010**
Whole milk	497 ml	432 ml
Skimmed milk	1.13 litres	1.15 litres
Cream	22 ml	19 ml

16 Jacob takes a maths test and a science test in December. He takes another test in each subject the following June. The table shows his scores.
Paul says "I improved most in science".
His teacher says "Actually you improved most in maths".

How did Paul decide? How did his teacher decide?

You **must** shows your working to justify your answer.

	December	June
Maths	46%	62%
Science	53%	71%

Learn... 4.3 Using ratios and proportion

Ratios are a good way of comparing quantities. The quantities must be in the same units.

Two (or more) ratios that simplify to the same ratio are called equivalent ratios.

For example, $8:12$ and $100:150$ are equivalent because they both simplify to $2:3$.

To simplify a ratio, divide each part by the same number.

> £3 = 300 pence
> When the amounts are in the **same units**, you can omit the units.

For example, the ratio of £3 to 40 pence = $300:40 = 30:4 = 15:2$

Each number is divided by 10, then 2

> You can use the fraction key on your calculator to simplify ratios.

This is the **simplest form** of this ratio. It uses the smallest possible whole numbers.

Sometimes ratios are divided further until one side is 1. This gives the $1:n$ (or the $n:1$) form. These are called **unitary ratios**.

> Dividing the original ratio $300:40$ by 40 also gives this.

Dividing $15:2$ by 2 gives the ratio $7\frac{1}{2}:1$ or $7.5:1$

The scales of maps and models are often given as unitary ratios.

The **unitary method** is based on working out what happens with **one** unit of something. It can be used to solve a variety of different problems involving ratio and **proportion**.

There are some links between ratios and fractions.

For example, suppose a brother and sister share an inheritance in the ratio $3:4$
- This means that for every £3 the brother gets, the sister gets £4.
- The brother's share is $\frac{3}{4}$ of the sister's share. The sister's share is $\frac{4}{3}$ or $1\frac{1}{3}$ times the brother's share.
- The brother gets $\frac{3}{7}$ of the whole inheritance and the sister gets the other $\frac{4}{7}$

The **multiplier method** multiplies the quantity by the fraction representing the ratio.

Example: A model of a car is made using a scale of 1 : 50

    a   The model car is 9 centimetres long.
        How long is the real car in metres?

    b   The real car is 2 metres wide. How wide is the model car in centimetres?

Solution: **a**   The ratio 1 : 50 means that 1 cm on the model car represents 50 cm on the real car.

        To find the length of the real car, multiply the length of the model car by 50.

        Length of the real car = 9 cm × 50 = 450 cm

        Length of the real car in metres = 450 ÷ 100 = 4.5 m

> 1 metre = 100 centimetres
> You can change the units before or after using the scale.

    b   The width of the real car = 2 m = 200 cm

        To find the width of the model car, divide the width of the real car by 50.

        Width of model = 200 ÷ 50 = 4 cm

Example: Liam earns £97.20 for working 15 hours in a supermarket.

How much does he earn for 24 hours at the same rate of pay?

> The amount Liam earns is **proportional to** the time he works. (If he works twice as long, he gets paid twice as much. If he works 3 times as long, he gets paid 3 times as much … and so on.)

Solution: For 15 hours Liam earns £97.20.

Unitary method

For 1 hour he earns £97.20 ÷ 15 = £6.48   Divide the pay for 15 hours by 15.

For 24 hours he earns £6.48 × 24 = £155.52   Multiply the pay for 1 hour by 24.

Always check that your answer looks reasonable.
Here the pay for 24 hours is more than that for 15 hours. ✓

Multiplier method

$£97.20 \times \frac{24}{15} = £155.52$

Multiplying by $\frac{24}{15}$ does the same as dividing by 15 and multiplying by 24.

> **Bump up your grade**
>
> To get a Grade C you need to be able to use the unitary method to solve ratio and proportion problems.

Example: **a**   Which jar of coffee gives the best value for money?

    b   Give a reason why you might decide to buy one of the other jars.

Solution: You can use the **unitary method** to solve 'best buy' problems.

    a   Find the cost of **1 gram** in each jar.
        Working in pence gives easier numbers to compare.

Small jar: Cost of 50 g = 156 pence
                Cost of 1 g   = 156p ÷ 50
                             = 3.12 pence

> There are sometimes other methods you could use.
> In this problem you could compare the cost of 100 g.

Medium jar: Cost of 100 g = 229 pence
                Cost of 1 g   = 229p ÷ 100
                             = 2.29 pence

> Small jar: 100 g (2 jars) costs 156p × 2 = 312 pence
> Medium jar: 100 g costs 229 pence

Large jar: Cost of 200 g = 445 pence
                Cost of 1 g   = 445p ÷ 200
                             = 2.225 pence

> Large jar: 100 g ($\frac{1}{2}$ jar) costs 445p ÷ 2 = 222.5 pence
> This also shows the large jar gives the best value for money

The cost of 1 gram of coffee is **least** in the large jar so the large jar gives the best value for money.

    b   You might buy a smaller jar if you only want a small amount of coffee (or if you do not have £4.45 to spend).

Example: Here is a recipe for blackberry and apple jam.

Sharon has picked 500 g of blackberries.

What quantity of apples and sugar does she need to use?

> Recipe: Blackberry and Apple Jam
> * 800 g Blackberries
> * 600 g Apples
> * 1kg Sugar

Solution: Writing the quantity of sugar as 1000 g gives the ratio of blackberries : apples : sugar
= 800 g : 600 g : 1000 g

You can change the quantity of blackberries from 800 g to 500 g in two steps.
Dividing by 8 reduces the quantity to 100 g. Then multiplying by 5 increases it to 500 g.

Blackberries	**Apples**	**Sugar**
800 g	600 g	1000 g
100 g	75 g	125 g
500 g	375 g	625 g

÷8, ×5 (Blackberries); ÷8, ×5 (Apples); ÷8, ×5 (Sugar)

Alternatively
You could divide by 200 to give the simplest form 4 : 3 : 5
Then to change the 4 into 500, you need to multiply by 125.
Doing this to all parts gives 500 : 375 : 625

If you prefer, you can divide by 800 to reduce the quantity of blackberries to **1 g** (the unitary method), then multiply by 500. This also gives the correct answer but the numbers are a bit harder.

In ratio and proportion questions you can multiply or divide by anything you like. But you must do the same to all the parts or quantities.

> AQA / *Examiner's tip*
> Write all parts in the same units before simplifying a ratio.

Practise...   4.3   Using ratios and proportion   

D

1
   a   Write down three different pairs of numbers that are in the ratio 1 : 3

   b   Explain how you can tell that two numbers are in the ratio 1 : 3

2
   Each of these ratios simplifies to 1 : 7
Copy these ratios and fill in the gaps.

   a   2 : __   **c**   __ : 21   **e** a : __

   b   5 : __   **d**   __ : 3500

3
   James is making fruit juice, using fruit cordial and water.

The label on the cordial says "To make 1 litre of fruit juice, use 200 ml of cordial and 800 ml of water."

James uses 250 ml cordial and 850 ml water.

Is James using the correct proportions?

Show working to justify your answer.

4
   A builder makes mortar by mixing cement and sand in the ratio 1 : 5

   a   How many buckets of sand does he need to mix with 3 buckets of cement?

   b   How many buckets of cement does he need to mix with 10 buckets of sand?

   c   How many buckets of cement and sand does he need to make 30 buckets of mortar?

D

5 The numbers *x* and *y* are in the ratio 3 : 4

   a If *x* is 12, what is *y*?

   b If *y* is 12, what is *x*?

   c If *x* is 1, what is *y*?

   d If *y* is 1, what is *x*?

   e If *x* and *y* add up to 35, what are *x* and *y*?

6 Here is a list of the ingredients you need to make 20 peanut cookies.

   a List the ingredients you would need to make:

      i 10 cookies

      ii 50 cookies.

   b Ewan wants to make some cookies.
He has bought a 200 g bag of peanuts.

      i How many peanut cookies can he make with this bag of peanuts?

      ii How much of each of the other ingredients will he need?

Recipe: Peanut Cookies (makes 20)
* 100 g Butter
* 50 g Sugar
* 150 g Flour
* 50 g Peanuts

7 A shop sells multi-packs of batteries.
A pack of 20 batteries costs £4.95. A pack of 32 batteries costs £6.28.
Which pack gives the best value for money?

C

8 Nina is paid £87.50 for 14 hours' work.

   a How much does she get paid for 20 hours' work?

   b She is paid £100. How many hours has she worked?

   c What assumption do you have to make to answer these questions?

9 Amy and Bianca go to Paris for the weekend.

   a They buy some euro at a bank. Amy gets 300 euro for £250.

      i What is the exchange rate in euro per £?

      ii How many euro does Bianca get for £275?

   b When they return, they go back to the bank to sell the euro they have left.
Amy gets £40 for 50 euro.

      i How much does Amy get for each euro?

      ii How much does Bianca get for 105 euro?

10 5 miles is approximately equal to 8 kilometres.

   a The distance from Southampton to Sheffield is 195 miles.
How far is this in kilometres?

   b The distance from Barcelona to Madrid is 624 kilometres.
How far is this in miles?

11 A box of chocolates contains milk chocolates, plain chocolates and white chocolates in the ratio 4 : 3 : 2

   a What fraction of the chocolates is:

      i milk

      ii plain

      iii white?

   b Show how you can check your answers to part **a**.

C

12 The ratio of men to women on a holiday cruise is $3 : 5$
What percentage of the people on the cruise are women?

⚠ 13 The weight of a pile of textbooks is proportional to the number of textbooks in the pile.

a Copy and complete this table.

Number of books	0	10	20	30
Weight (kg)		3.25		

b Plot the values in the table as points on a graph, using the number of books as the *x*-coordinates and their weight in kilograms as *y*-coordinates.

c The points should lie in a straight line through (0, 0).

i Explain why.

ii Use the graph to find the weight of a pile of 24 books.

iii Mr Marks says he cannot lift anything heavier than 5 kilograms. How many books can he carry?

⚠ 14 The cost of calls on a mobile phone is proportional to the length of the calls in minutes.

The cost per minute depends on whether calls are in peak or off-peak periods.

This graph shows the cost of calls made in peak periods.

a **i** Make a copy of the graph on graph paper.

ii In off-peak periods the cost of a call is 15 pence per minute.

iii Draw a line on your graph to show the cost of off-peak calls.

b **i** Tom makes a peak period call that lasts half an hour. How much does it cost?

ii How much would Tom have saved if he had made the call in an off-peak period?

iii Bill spends £12 on a peak period call. How much would the same length call have cost in an off-peak period?

c A contract phone costs £15 per month. You get 60 minutes of 'free' calls and pay for all other calls at 20 pence per minute.

i Draw a new graph to show this.

ii Bill uses a contract phone. Is the cost of calls proportional to the length of the calls? Give a reason for your answer.

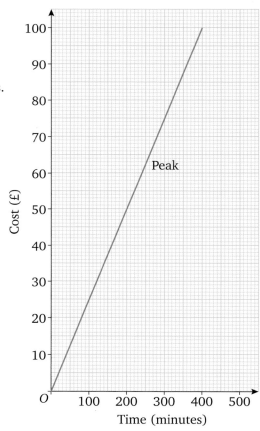

15 The scale on a map is $1 : 50\,000$

a The distance between two landmarks on the map is 12 centimetres. Find the actual distance.

b Kathy says the distance between these landmarks on a map with a scale of $1 : 25\,000$ will be 6 cm.

Is she correct? Give a reason for your answer.

16

Road signs use ratios or percentages to give the gradients of hills.

Ratio of vertical
distance : horizontal distance

Vertical distance as a
percentage of the horizontal distance

Which of the hills described by these road signs is steeper? Give a reason for your answer.

17   When you enlarge a photograph, the ratio of the height to width must stay the same.
If the ratio is different the objects in the photograph will look stretched or squashed.

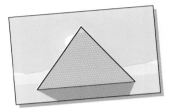 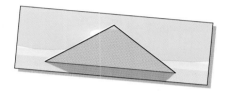

a   Chloe has a photo of her favourite group that is 15 cm wide and 10 cm high.
What is the ratio of width to height in its simplest form?

b   Chloe wants to put an enlarged copy of the photo
in a frame on her bedroom wall.
The table gives the sizes of the frames she can buy.
Which of these frames is most suitable?
Give a reason for your answer.

Frame	Width (cm)	Height (cm)
A	25	20
B	30	25
C	40	30
D	45	30
E	50	40

18   The table on the right gives the ages of the children
who are booked in to a nursery.

The nursery has three rooms: one for children under
2 years old, one for two-year-olds and one for children
aged 3 years and over.

The minimum adult : child ratios for nurseries are
given in the table below.

Age (years)	Number of children	
	morning	afternoon
0	2	1
1	4	2
2	6	9
3	5	8
4	2	1

Age	Minimum adult : child ratio
Children under 2 years	1 : 3
Children aged 2 years	1 : 4
Children aged 3–7 years	1 : 8

How many staff does the nursery need:

a   in the morning   **b**   in the afternoon?

19   **a**   Which size of shampoo bottle gives
the best value for money?
You **must** show all your working and
give a reason for your answer.

b   Why might someone buy a
different size?

20 Students at a school can visit a theme park, a zoo or a safari park.
The table shows how many have chosen each place.

Choice	Number of students
Theme park	124
Zoo	76
Safari Park	98

The school's policy is to have a maximum child : adult ratio of 8 : 1 on school visits.

There are 20 teachers available and some parents have offered to go on the visits if they are needed.

How many parents are needed?

4 Assess (k!)

F

1 This diagram is made from equilateral triangles.

a What percentage of the diagram is:

i shaded **ii** not shaded?

b Another diagram has 65% shaded.
What fraction of the diagram is shaded?
Simplify your answer.

2 Molly got $\frac{5}{8}$ of the marks in a test. Rose got 65%.

Who did better in the test? Give a reason for your answer.

E

3 A painter has two tins of paint.

The 5 litre tin, A, is 35% full of paint.
The 2.5 litre tin, B, is two-thirds full of paint.

Which tin contains more paint? You **must** show your working.

4 A Sat Nav usually costs £159.99.
It is reduced by 15% in a sale.

What is the cost of the Sat Nav in the sale?

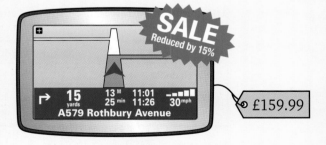

D

5 A school buys 10 bottles of milk for drinks at a parents' evening.
Each bottle holds enough for 25 drinks.
They make 235 drinks.

Calculate the percentage of milk used.

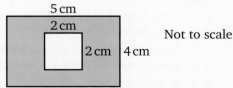

6 What percentage of this shape is shaded blue?

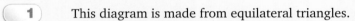

5 cm
2 cm
2 cm 4 cm
Not to scale

7 168 men and 210 women book a holiday cruise.

Write the ratio of the number of men to the number of women:

a in its simplest form **b** in the form 1 : n

D

8 The sizes of the interior angles of a quadrilateral are in the ratio $2 : 2 : 3 : 5$
The equal angles are both 60°. Calculate the size of the other angles.

9 The table shows the amounts needed to make 24 fruit biscuits.

Ingredient	Amount for 24 biscuits (g)
Flour	300
Butter	150
Sugar	100
Fruit	120

Calculate the amounts needed to make 36 fruit biscuits.

10 In the 1908 Olympic Games, Reggie Walker won the 100 metres in a time of 10.8 seconds.
A century later, in the 2008 Olympics, Usain Bolt won the 100 metres in 9.69 seconds.
Calculate the percentage decrease in the winning time.

C

11 It costs £259.80 for 20 square metres of carpet.
How much does it cost for 36 square metres of the same carpet?

12 Sun tan lotion is sold in two different sizes: small and large.

a Which bottle gives the better value for money?

b Give one reason why you might prefer to buy the other bottle.

AQA Examination-style questions 🔑

1 **a** Travelling 10 000 kilometres costs £800 for petrol.
How much does it cost to travel 12 500 kilometres? *(1 mark)*

b Last year Mr Taylor travelled 15 000 km in his car and spent £1200 on petrol.
This year he expects to travel 20 000 km.
He estimates that the price of petrol has increased by 10% on what it was last year.
How much should Mr Taylor expect to pay for petrol this year? *(4 marks)*

AQA 2009

2 **a** A water meter at a house records the volume of water used, in cubic metres.
The meter readings at the start and end of a 3-month period are as follows.

	Reading in cubic metres
End	4205
Start	4154

Water costs 104p per cubic metre.
Find the cost of the water used in this period.
Give your answer in pounds. *(4 marks)*

b In this period the cost of water at another house is £62.
The sewage charge is 97% of the cost of the water.
Find the sewage charge. *(2 marks)*

c A factory uses 34 cubic metres of water one week and 39 cubic metres in the following week.
Calculate the percentage increase in the consumption of water. *(3 marks)*

AQA 2008

5 Perimeter and area

Objectives

Examiners would normally expect students who get these grades to be able to:

G

find the perimeter of a shape by counting sides of squares

find the area of a shape by counting squares

estimate the area of an irregular shape by counting squares and part squares

name the parts of a circle

F

work out the area and perimeter of a simple rectangle, such as 5 m by 4 m

E

work out the area and perimeter of a harder rectangle, such as 2.6 cm by 8.3 cm

D

find the area of a triangle and parallelogram

find the area and perimeter of shapes made from triangles and rectangles

calculate the circumference and area of a circle

C

work out the perimeter and area of a semicircle.

Key terms

perimeter	diameter
area	segment
base	tangent
perpendicular height	arc
circle	sector
circumference	chord
radius, radii	

Did you know?

Surface area of the Earth

The Earth looks blue from space because about 70% of the Earth's surface area is water. Some of the larger countries in the world such as Russia, China and the USA take up a lot of the 30% of the Earth's surface area which is land.

Google Earth can be used to find where you live. Did you know that you can also use Google Earth to measure the perimeter and area of your house or your school grounds? Try it!

You should already know:

✓ how to add, subtract, multiply and divide whole numbers and decimals

✓ definitions of square, rectangle and triangle.

Learn... 5.1 Perimeter and area of a rectangle

The distance round the outside of a shape is called the **perimeter**.

This rectangle is 6 m long and 4 m wide.

To find the perimeter, imagine walking all the way round the edge.

The distance all the way round is **6 m + 4 m + 6 m + 4 m = 20 m**

So the perimeter of the rectangle is 20 m.

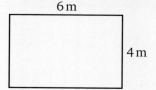

The **area** of a shape is the amount of space it covers.

Area is measured in square units.

The rectangle has been drawn on squared paper.

Counting the squares covered by the rectangle gives an area of 24 square units. As the measurements are in metres the area is 24 square metres or 24 m².

Counting squares can take a long time. In this rectangle, each row has 6 squares (the length). There are 4 rows (the width) so the total number of squares is $6 \times 4 = 24$

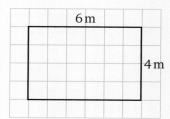

So:

Area of rectangle = length × width

Example: Find the perimeter and area of each shape.

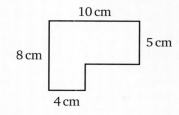

Solution: **a**   Perimeter = 3.5 + 8.2 + 3.5 + 8.2 = 23.4 cm

Area = 3.5 × 8.2 = 28.7 cm²

b   Perimeter method 1: To find the perimeter of this shape, first work out the missing lengths.

Missing width: 10 cm − 4 cm = 6 cm

Missing height: 8 cm − 5 cm = 3 cm

So the total perimeter = 10 + 5 + 6 + 3 + 4 + 8 = 36 cm

> **AQA Examiner's tip**
>
> Mark the corner you are starting from and imagine walking right round the shape. This will ensure that you don't miss out any sides.

Perimeter method 2: With L shapes, the perimeter is the same as the perimeter of a rectangle with length and width equal to the longest sides. Using this fact, you can find the perimeter without working out the missing lengths.

Perimeter = 10 + 8 + 10 + 8 = 36 cm

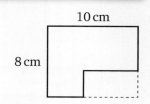

Area: To work out the area, divide the shape into rectangles, for example, A and B as shown.

Area of rectangle A = 10 × 5 = 50 cm²

Area of rectangle B = 4 × 3 = 12 cm²

Total area of shape = 50 + 12 = 62 cm²

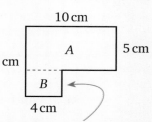

This length is 8 cm − 5 cm = 3 cm

> **AQA Examiner's tip**
>
> There is often more than one way to divide up a shape into rectangles.

Practise...

5.1 Perimeter and area of a rectangle

G F E D C

G

1 These diagrams are drawn on centimetre squared paper.

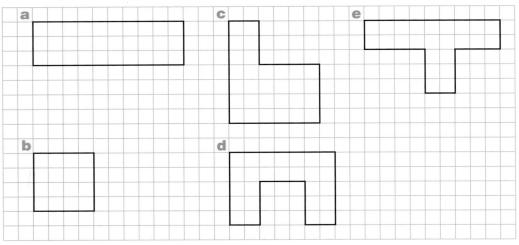

i   Work out the perimeter of each shape.

ii   Work out the area of each shape by counting squares.

2 Estimate the area of this shape.

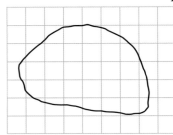

Hint

If more than half a square lies inside the shape, count it as a whole square.
If less than half the square lies inside the shape do **not** count it.

AQA *Examiner's tip*

On your copy, number the squares as you count them.

3 Look at these four rectangles.

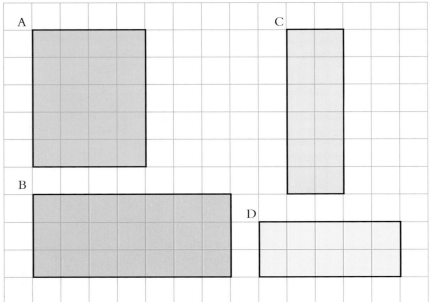

a   Which rectangle has the largest perimeter?

b   Which rectangle has the smallest perimeter?

c   Which rectangle has the largest area?

d   List the rectangles and their areas in decreasing order of size.

4 Work out the perimeter of squares with the following side lengths.

a 12 cm

b 20 cm

c 4.6 cm

d 13.2 cm

e 126.5 cm

5 On centimetre squared paper draw two different rectangles with a perimeter of 12 cm.

6 For each of these shapes, work out:

i the perimeter

ii the area.

a 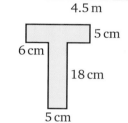 2.7 cm

8.3 cm

b 1.4 cm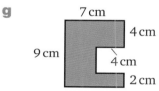

8.2 cm

c

6 m

d

5 cm

e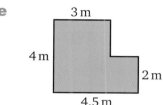

3 m

4 m

2 m

4.5 m

Not drawn accurately

f 5 cm

6 cm

18 cm

5 cm

g 7 cm

4 cm

9 cm

4 cm

2 cm

h 12 m

8 m

4 m

1 m

7 Find the area of rectangles with these measurements.

a Length 3.7 cm, width 28 mm

b Length 2.1 m, width 84 cm

Hint

Be aware of the units of length when calculating area and perimeter.

8 A rectangle of length 40 cm and width 19.6 cm has the same area as a square.
Work out the length of the square.

9 A rug is 1.5 m long and 0.8 m wide. It is laid on the floor of a room which is 3.2 m long and 3 m wide.
Work out the percentage of the floor area that the rug will cover.

10 Andy and Katie would like name plaques for their bedroom doors.

The carpenter makes the letters by cutting and sticking together two types of rectangular pieces of wood.

Piece A – length 8 cm, width 4 cm   Piece B – length 6 cm, width 4 cm

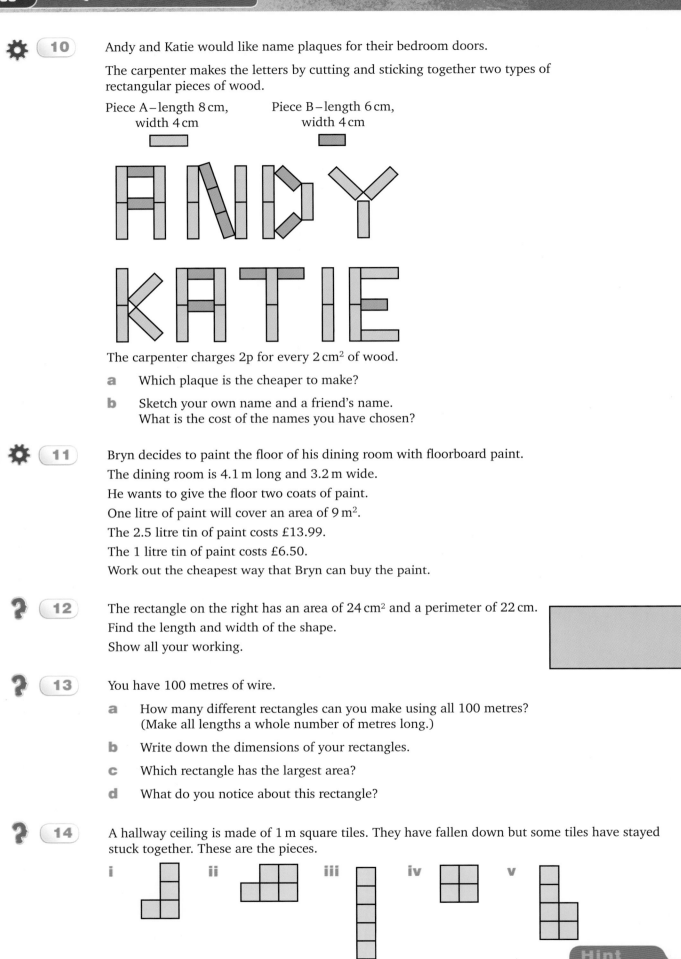

The carpenter charges 2p for every 2 cm² of wood.

a Which plaque is the cheaper to make?

b Sketch your own name and a friend's name.
What is the cost of the names you have chosen?

11 Bryn decides to paint the floor of his dining room with floorboard paint.
The dining room is 4.1 m long and 3.2 m wide.
He wants to give the floor two coats of paint.
One litre of paint will cover an area of 9 m².
The 2.5 litre tin of paint costs £13.99.
The 1 litre tin of paint costs £6.50.
Work out the cheapest way that Bryn can buy the paint.

12 The rectangle on the right has an area of 24 cm² and a perimeter of 22 cm.
Find the length and width of the shape.
Show all your working.

13 You have 100 metres of wire.

a How many different rectangles can you make using all 100 metres?
(Make all lengths a whole number of metres long.)

b Write down the dimensions of your rectangles.

c Which rectangle has the largest area?

d What do you notice about this rectangle?

14 A hallway ceiling is made of 1 m square tiles. They have fallen down but some tiles have stayed stuck together. These are the pieces.

i   **ii**   **iii**   **iv**   **v**

a Sketch out the tiles and put them together to fit back on the ceiling.

b What is the area of the ceiling?

Hint
The ceiling is rectangular.

 5.2 **Area of parallelograms and triangles**

Area of a parallelogram

Here is a parallelogram drawn on squared paper.

If the shaded triangle is cut from one end of the parallelogram and put on to the other end, the shape becomes a rectangle.

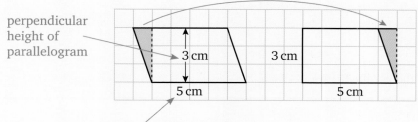

perpendicular height of parallelogram

base of parallelogram

The area of the parallelogram is the same as the area of the rectangle.

The **base** of the parallelogram is the same as the length of the rectangle.

The **perpendicular height** of the parallelogram is the same as the width of the rectangle.

Area of rectangle $= 5 \times 3 = 15\,cm^2$

Area of parallelogram $= 5 \times 3 = 15\,cm^2$

The area of any parallelogram can be calculated using the formula:

Area of parallelogram = base × perpendicular height

Area of a triangle

If you draw a diagonal on the parallelogram, the parallelogram is divided into two equal triangles.

The area of each triangle is half the area of the parallelogram.

This gives the formula for the area of a triangle:

Area of triangle $= \frac{1}{2} \times$ base × perpendicular height

If a rectangle is divided into two triangles, then:

- the base of each triangle is the length of the rectangle
- the perpendicular height of each triangle is the width of the rectangle.

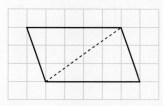

Example: Calculate the area of each triangle.

a

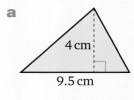

4 cm

9.5 cm

b

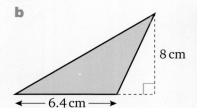

8 cm

6.4 cm

 *Examiner's tip*

Be careful to use the perpendicular height and not the length of the sloping sides of the triangle.

Solution: **a**   From the diagram: base = 9.5 cm   perpendicular height = 4 cm

So area $= \frac{1}{2} \times$ base × perpendicular height

$= \frac{1}{2} \times 9.5 \times 4 = 19\,cm^2$

b   From the diagram: base = 6.4 cm   perpendicular height = 8 cm

So area $= \frac{1}{2} \times$ base × perpendicular height

$= \frac{1}{2} \times 6.4 \times 8 = 25.6\,cm^2$

Example: A parallelogram has an area of 51 cm².

The perpendicular height of the parallelogram is 3 cm.

Work out the base of the parallelogram.

Solution: Using the formula, area = base × perpendicular height

51 = base × 3   Divide each side by 3.

$\frac{51}{3}$ = base

17 = base

The base of the parallelogram is 17 cm.

Practise... 5.2 Area of parallelograms and triangles

G F E D C

D

1   Work out the area of each parallelogram.

a
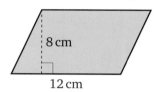
8 cm
12 cm

c
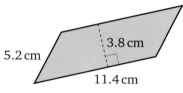
5.2 cm
3.8 cm
11.4 cm

b
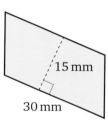
15 mm
30 mm

d
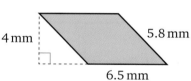
4 mm
5.8 mm
6.5 mm

2   Work out the area of each triangle.

a
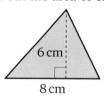
6 cm
8 cm

c
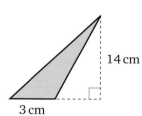
14 cm
3 cm

b
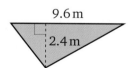
9.6 m
2.4 m

d

18 cm
5 cm

3   Which of the following shapes has the largest area? Show your working.

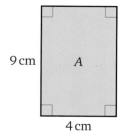
9 cm   A
4 cm

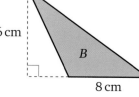
6 cm
B
8 cm

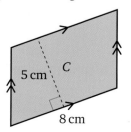
5 cm   C
8 cm

D

4   A parallelogram of base 7.2 cm has an area of 97.2 cm².

Work out the height of the parallelogram.

5   Four students are trying to find the area of this triangle.

Kieran thinks the answer is 120 cm² because $8 \times 15 = 120$

Javed thinks the answer is 40 cm² because $8 + 15 + 17 = 40$

Leanne thinks the answer is 60 cm² because $\frac{1}{2} \times 15 \times 8 = 60$

Megan thinks the answer is 127.5 cm² because $\frac{1}{2} \times 15 \times 17 = 127.5$

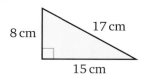

a   Who is correct?

b   What mistakes have the other students made?

6   Work out the perimeter and the area of these parallelograms.

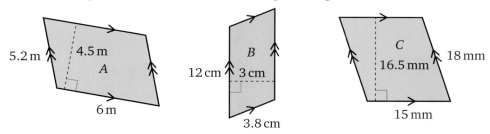

7   A triangle has a perpendicular height of 22 cm.

The area of the triangle is 308 cm².

Work out the length of the base of the triangle.

⚠ **8**   Work out the area of each of these shapes.

a

b

⚙ **9**   This club logo is to be made out of two metal triangles.

a   Linda works out the dimensions of the smallest rectangle of metal from which the logo can be cut out are 10 cm by 14 cm.

Find a smaller rectangle from which the logo can be cut out.

b   **i**   What percentage of metal is wasted from Linda's rectangle?

ii   What percentage of metal is wasted from the smallest rectangle?

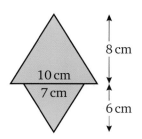

⚙ **10**   A company makes kites.

They cut triangles from yellow silk and blue silk as shown.

The yellow silk costs £5 per square metre and the blue silk costs £7 per square metre.

Assuming that the triangles for several kites can be cut from the material without wastage, find the cost of material used for each kite.

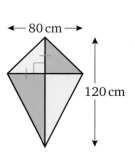

Learn... 5.3 Compound shapes

To find the area of a compound shape, divide up the shape into rectangles, parallelograms and triangles.

You can divide this shape into two triangles and a rectangle and work out the area of each smaller shape.

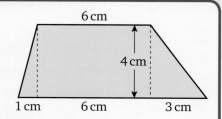

Total area =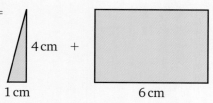

$$\text{area} = \tfrac{1}{2} \times 1 \times 4 \qquad \text{area} = 6 \times 4 \qquad \text{area} = \tfrac{1}{2} \times 3 \times 4$$
$$= 2\,\text{cm}^2 \qquad\qquad = 24\,\text{cm}^2 \qquad\qquad = 6\,\text{cm}^2$$

Total area = 2 + 24 + 6 = 32 cm²

The shape in this Learn is a trapezium.

The area of a trapezium can also be found by using the formula:

$$\text{area of trapezium} = \tfrac{1}{2}(a + b)h$$

or in words, 'Add together the parallel sides and multiply by half the distance between them.'

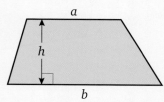

You should try to use the formula for area of a trapezium whenever a compound shape includes a trapezium.

Example: Work out the perimeter and the area of this trapezium.

> **AQA Examiner's tip**
> The formula for a trapezium is given in the formula sheet for Unit 3.

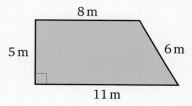

Solution: Perimeter = 8 + 6 + 11 + 5 = 30 m

$$\text{Area} = \tfrac{1}{2}(a + b)h = \tfrac{1}{2}(8 + 11) \times 5 = \tfrac{1}{2} \times 19 \times 5 = 47.5\,\text{m}^2$$

Example: Work out the area of this shape.

> **AQA Examiner's tip**
> Remember to state the units in your answers if the question asks for it.

Solution: The shape can be split into a trapezium and a triangle.

The area of the triangle is $\tfrac{1}{2} \times$ base $\times$ height $= \tfrac{1}{2} \times 35 \times 26$

$$= 455\,\text{mm}^2$$

The area of the trapezium is $\tfrac{1}{2}(a + b)h = \tfrac{1}{2}(35 + 30) \times 18$

$$= \tfrac{1}{2} \times 65 \times 18$$

$$= 585\,\text{mm}^2$$

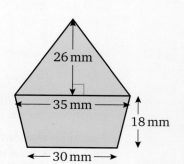

The total area of the shape is 455 + 585 = 1040 mm²

Practise... 5.3 Compound shapes

D

1 Each of these shapes is a trapezium.

Work out the area of each shape.

a
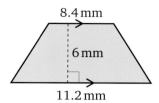
8.4 mm
6 mm
11.2 mm

c
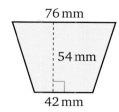
76 mm
54 mm
42 mm

b
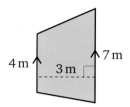
4 m
3 m
7 m

d
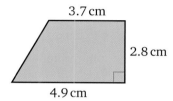
3.7 cm
2.8 cm
4.9 cm

2 Work out the area of each shape.

a
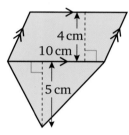
4 cm
10 cm
5 cm

c
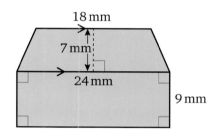
18 mm
7 mm
24 mm
9 mm

b
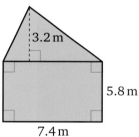
3.2 m
5.8 m
7.4 m

d
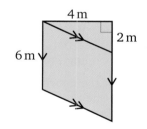
4 m
2 m
6 m

3 For each of these shapes, find:
  i the perimeter
  ii the area.

a
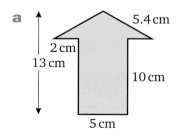
5.4 cm
2 cm
13 cm
10 cm
5 cm

c
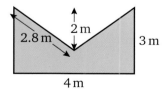
2 m
2.8 m
3 m
4 m

Not drawn accurately

e
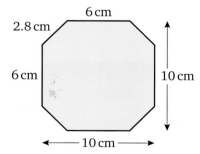
6 cm
2.8 cm
6 cm
10 cm
10 cm

b
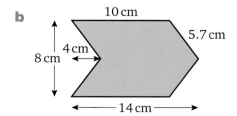
10 cm
5.7 cm
8 cm
4 cm
14 cm

d
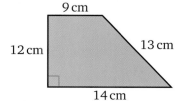
9 cm
12 cm
13 cm
14 cm

D
C

4   **a**   Work out the area of the shaded part of each shape.

i

14 cm
12 cm   25 cm
16 cm
20 cm

ii

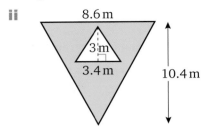

8.6 m
3 m
3.4 m
10.4 m

b   What fraction of diagram **i** is shaded?

⚠ 5   Sam is making badges using the design shown. He uses silver foil for the surround of the white centre shape.

Sam has one more badge to make. He has 10 cm² of silver foil left.

Does Sam have enough foil to complete the badge?

Show your working.

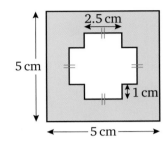

2.5 cm
5 cm
1 cm
5 cm

⚙ 6   The diagram shows Joe's allotment which is in the shape of a trapezium.

a   Joe wants to build a new fence to go round the allotment leaving the gate in the same place. The fencing will cost Joe £9.60 per metre. Work out the cost of the new fencing.

b   Joe wants to plant vegetables in $\frac{3}{4}$ of his allotment. Work out the area of his allotment which will be used for vegetables.

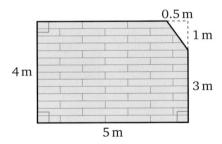

42 m
38 m   34 m
gate
2.8 m
42 m

⚙ 7   This is a sketch of Vikki's dining room.

a   Vikki wants to paint the floorboards which cover the whole of the floor. One tin of floorboard paint will cover 9.5 m² of floor. Vikki thinks that two tins will cover all her floor.

Is she correct?

You must show your working.

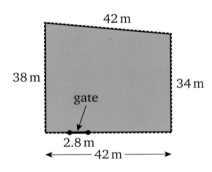

0.5 m
1 m
4 m
3 m
5 m

b   Vikki fits new skirting board all round the room except across the diagonal corner entrance. Skirting board is sold in 2.4 m lengths costing £7.99 each.

Work out the total cost of the skirting board.

Learn...   **5.4 Circumference and area of a circle** 🄺

In Learns 5.1, 5.2 and 5.3 you found perimeters and areas of various shapes. In this Learn you will find the perimeter (or **circumference**) and area of a **circle**.

First you will need to know the names of various parts of a circle.

circumference: the distance all the way round the circle

diameter: the distance from one side of the circle to the other, through the centre

radius: the distance from the centre of the circle to the circumference

Other parts of the circle are:

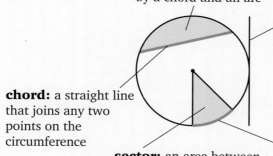

segment: an area enclosed by a chord and an arc

tangent: a straight line outside the circle that touches the circle at only one point

chord: a straight line that joins any two points on the circumference

arc: a section of the circumference

sector: an area between two radii and an arc

The importance of pi

π is the Greek letter pi and represents a value of 3.14159... (Look for the π button on your calculator.)

Circumference of a circle

The circumference, C, of a circle can be calculated using the formula:

$C = \pi d$ where d is the diameter of the circle   Remember that $d = 2 \times r$

Area of a circle

The area, A, of a circle with radius r can be calculated using the formula:

$A = \pi r^2$

Example: Find the circumference and area of a circle of diameter 12 cm.

Give your answers to one decimal place.

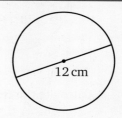

12 cm

Solution: $C = \pi d$ so $C = \pi \times 12$

Using the π button on a calculator gives $C = 37.69911...$

So the circumference of the circle is 37.7 cm (to 1 d.p.).

To find the area we must use the radius. If the diameter is 12 cm the radius is $\frac{12}{2} = 6$ cm.

$A = \pi r^2$ so $A = \pi \times 6^2$ or $\pi \times 6 \times 6$

Using the π button on a calculator gives $A = 113.0973...$

So the area of the circle is 113.1 cm² (to 1 d.p.).

> AQA *Examiner's tip*
>
> Always state the units of your answer and remember that area is always measured in square units.

Example: Find the diameter of a circle of circumference 22.3 cm.

Give your answer to one decimal place.

Solution: $C = \pi d$ so $22.3 = \pi \times d$

Divide both sides by π.

$d = \dfrac{22.3}{\pi} = 7.0983...$

The diameter is 7.1 cm (to 1 d.p.).

5.4 Circumference and area of a circle

Practise...

G F E D C

G

1 Name the parts of a circle labelled *A–D* on this diagram.

O is the centre of the circle.

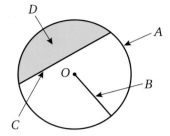

2 **a** Give the name for a line drawn from one side of a circle to the other, passing through the circle's centre.

b Give the name for the region between two radii and an arc of a circle.

c Give the name for a straight line outside a circle which touches the circle at only one point.

D

3 Calculate the circumference of each circle.

Give each answer to one decimal place.

a 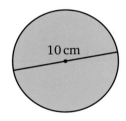 10 cm

b 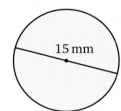 15 mm

c 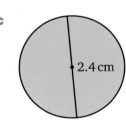 2.4 cm

d 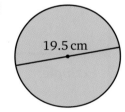 19.5 cm

4 Calculate the circumference of each circle.

Give each answer to one decimal place.

a 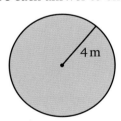 4 m

b 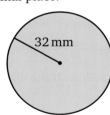 32 mm

c 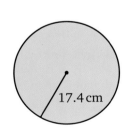 17.4 cm

d 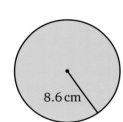 8.6 cm

5 A circle has a circumference of 62.8 cm.

Work out the diameter of this circle.

Give your answer to the nearest whole number.

6 Calculate the area of each circle.

Give each answer to two decimal places.

a 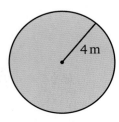 4 m

b 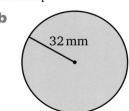 32 mm

c 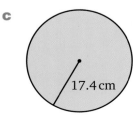 17.4 cm

d 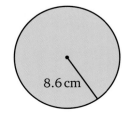 8.6 cm

7 Calculate the area of each circle.

Give each answer to two decimal places.

a 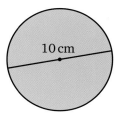 10 cm

b 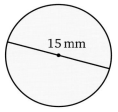 15 mm

c 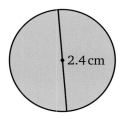 2.4 cm

d 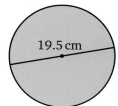 19.5 cm

8 For each semicircle, work out:

a the perimeter   **b** the area.

Give each answer to one decimal place.

i 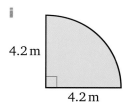 18 cm

ii 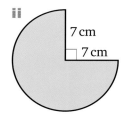 4.8 cm

Hint
The area of a semicircle is half the area of a circle having the same diameter or radius. The perimeter of a semicircle is half the circumference of a circle having the same diameter or radius **plus** the diameter.

Bump up your grade
To get a Grade C make sure you know how to work out the perimeter and area of a semicircle.

9 A circle has an area of 201 mm².

Calculate the radius of the circle giving your answer to the nearest whole number.

10 For each of these shapes, calculate:

a the perimeter   **b** the area.

i 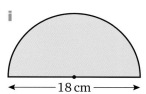 4.2 m, 4.2 m

ii 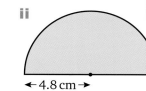 7 cm, 7 cm

11 Teri is training for a fun run. She wants to run 10 000 m each week during her training.
The diagram shows the running track where she trains.

Teri says, 'If I run five times round this track each day from Monday to Friday, I will have run more than 10 000 m in a week.'

Is she correct? Show your working.

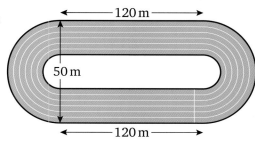 120 m, 50 m, 120 m

12 A circular cake tin has a diameter of 22.5 cm. The lid is sealed with tape. The ends of the tape overlap by 1.5 cm.

Calculate the length of tape needed to seal the tin.

13 The London Eye has a diameter of 135 m and takes approximately 30 minutes to complete one revolution. Passengers travel in capsules.
How far does the base of a capsule travel every 5 minutes?

5 Assess *k!*

F
E

1 Find the perimeter and the area of each of the following rectangles.

    a   Length 16 cm, width 12 cm

    b   Length 7.4 m, width 3.1 m

    c   Length 37 mm, width 15.2 mm

D

2 Find the area of each triangle.

a
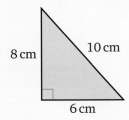

c
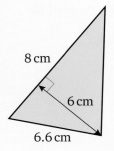

Not drawn accurately

b
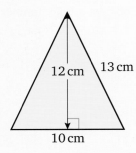

d
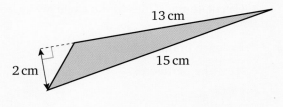

3 Find the area of each parallelogram.

a
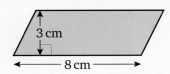

c

Not drawn accurately

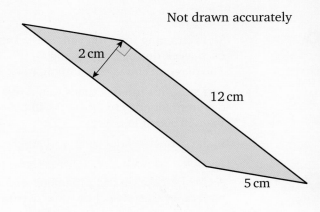

b
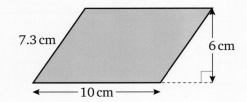

4 Find the circumference and the area of each circle.

a
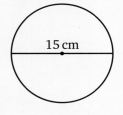
15 cm

b

14.6 mm

c
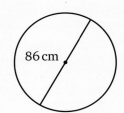
86 cm

d

33 cm

5 Work out the area of each of these shapes.

a

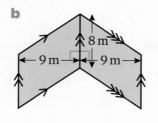

b

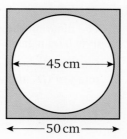

Not drawn accurately

6 A circle of diameter 45 cm is cut out of a square of side 50 cm.

Calculate the shaded area.

Not drawn accurately

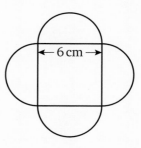

7 A shape is made from a square of side 6 cm surrounded by four semicircles of diameter 6 cm.

Work out the area of the shape.

Give your answer to one decimal place.

AQA Examination-style questions

1 The diagram shows a plan view of a landfill site on a centimetre grid.

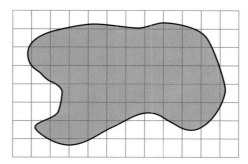

a Estimate the number of shaded squares in the diagram.
You **must** show your working.

(2 marks)

b A landscape gardener is going to cover the site with turf (grass).
The table shows the cost of turf for different areas (m²).

Area of turf (m²)	Cost per square metre
40–59	£2.83
60–130	£2.33
131–240	£2.03
241–480	£1.78
481–640	£1.53
641–960	£1.40
961–1440	£1.23

On the diagram, one square represents 4 m².
The landscape gardener must buy enough turf to cover the landfill site.
Work out how much he has to pay. You **must** show your working.

(3 marks)

AQA 2009

2   Three identical rectangles fit together as shown. Work out the total area.

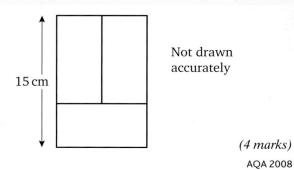

15 cm

Not drawn accurately

(4 marks)

AQA 2008

3   The diagram shows five shapes, *A*, *B*, *C*, *D* and *E*, drawn on a grid.

Put the shapes in order of area, starting with the smallest (*D*) and ending with the largest (*A*).

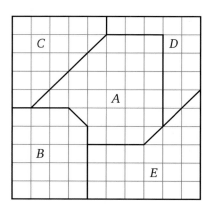

(2 marks)

AQA 2008

Objectives

Examiners would normally expect students who get these grades to be able to:

F

set up and solve a simple equation such as $5x = 10$ or $x + 4 = 7$

E

set up and solve an equation involving fractions such as $\frac{x}{3} = 4$ or $2x - 3 = 8$

D

set up and solve more complicated equations such as $3x + 2 = 6 - x$ or $4(2x - 1) = 20$

C

set up and solve an equation such as $4x + 5 = 3(x + 4)$ or $\frac{2x - 7}{4} = 1$

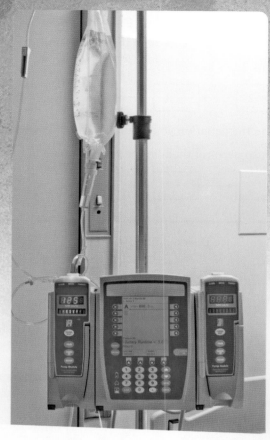

Did you know?

Nurses use algebra

Students often ask, 'Why do we have to do algebra at school when we are never going to use it again?'

This is not true. There are many jobs that involve algebra.

Nurses, for example, use algebra every day in their administration of medicines and drips.

Here is an equation that they would use to control an electronic drip for a patient.

$$\text{rate of drip (drops per min)} = \frac{\text{volume of infusion (ml)}}{\text{time (min)}}$$

Key terms

unknown
solution
operation
brackets
expanding
denominator
lowest common denominator

You should already know:

✔ the inverse operations of $+$, $-$, $\times$ and $\div$

✔ how to collect like terms

✔ how to use substitution

✔ how to multiply out brackets by a positive or negative number

✔ how to find the lowest common denominator for two fractions.

Learn... **6.1 Simple equations**

When you find the value for an **unknown**, you have found the **solution** to an equation.

Equations can involve any of the **operations** $+, -, \times, \div$.

Many equations involve more than one operation, e.g.

$2x + 3$ means 2 lots of x add 3.

$$x \longrightarrow \boxed{\times 2} \xrightarrow{2x} \boxed{+ 3} \longrightarrow 2x + 3$$

When solving the equation you would use the inverse of each operation.

You would also perform them in the reverse order.

$$x \longleftarrow \boxed{\div 2} \xleftarrow{2x} \boxed{- 3} \longleftarrow 2x + 3$$

Example: Solve the equation $6x = 24$

> **Hint**
>
> Remember that $6x$ means $6 \times x$

Solution:
$6x = 24$

$\dfrac{6x}{6} = \dfrac{24}{6}$ Divide both sides by 6.
      The inverse (opposite) of multiplying by 6 is dividing by 6.

$x = 4$

Check
$6 \times 4 = 24$ ✓

> AQA **Examiner's tip**
>
> Always check your answer by substituting its value back into the original equation.

Example: Solve the equation $x - 2 = 7$

Solution:
$x - 2 = 7$

$x - 2 + 2 = 7 + 2$ Add 2 to both sides. The inverse of subtracting 2 is adding 2.

$x = 9$

Check
$9 - 2 = 7$ ✓

Example: Solve the equation $4x + 3 = 17$

Solution: This is an equation with two operations: $\times$ and $+$

The expression $4x + 3$ was formed by multiplying by 4, then adding 3.

The inverse of adding 3 is subtracting 3. So to solve the equation you need to begin by subtracting 3 from both sides.

$4x + 3 - 3 = 17 - 3$ Subtract 3 from both sides.

$4x = 14$

$\dfrac{4x}{4} = \dfrac{14}{4}$ Divide both sides by 4.
      (Dividing by 4 is the inverse of multiplying by 4.)

$x = 3\frac{2}{4}$ Change all improper fractions to mixed numbers.

$x = 3\frac{1}{2}$ Simplify any fractions.

Check
$(4 \times 3\frac{1}{2}) + 3 = 17$ ✓

Example: The smallest angle of a triangle is $x°$.

The middle-sized angle is double the smallest angle plus $10°$.

The largest angle is double the middle-sized angle plus $10°$.

Calculate x.

Solution: The smallest angle $= x°$

The middle angle $= (2x + 10)°$

The largest angle $= 2(2x + 10)° + 10° = (4x + 20 + 10)°$

$x + (2x + 10) + (4x + 20 + 10) = 180$ The angles in a triangle add up to $180°$.

$7x + 40 = 180$ Collect like terms.

$7x + 40 - 40 = 180 - 40$ Subtract 40 from both sides.

$7x = 140$

$\dfrac{7x}{7} = \dfrac{140}{7}$ Divide both sides by 7.

$x = 20°$

Check

$20 + 2(20) + 10 + 4(20) + 20 + 10 = 20 + 40 + 10 + 80 + 20 + 10 = 180$ ✓

Practise...   6.1 Simple equations 🔊   G F E D C

1   Solve these equations.

   a $5x = 35$ **d** $y + 3 = 17$

   b $6b = 3.6$ **e** $c + 9.9 = 2.7$

   c $x - 3 = 8$ **f** $6 = 8 - 3t$

> **Hint**
>
> If the unknown is on the right-hand side of the equation, turn it round completely before you start to solve it, e.g.
>
> $4 = 13 - 6z$ becomes $13 - 6z = 4$

F

2   Set up and solve an equation to find the value of x in the diagram.

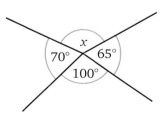

E

⚠ **3**   Two angles of a parallelogram are $2x + 40°$ and $3x$.
Find the value of x if:

   a   the angles are opposite each other

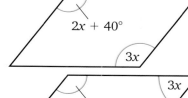

   b   the angles are next to each other.

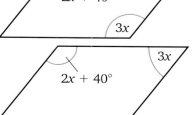

4 The diagram shows the position of two ships A and B.

The bearing of B from A is 110°.

Find the value of x.

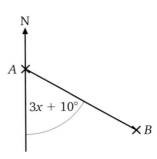

5 Nicole was given £2.00 to go to the shop and buy some cans of drink.

She bought two cans and was given 86p change.

a Write down an equation to represent this situation.

b Solve your equation to work out the cost of a can.

6 Simone thinks of a number, doubles it and subtracts 3. The answer is 7.

Use x to represent the number Simone thought of.

a Write down an equation in x using the information given.

b Solve your equation to find Simone's number.

7 The diagram shows three angles on a straight line.

a Write down an equation in x.

b Solve your equation to find the value of x.

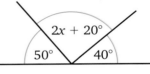

8 **a** Use the diagram to write down an equation in z.

b Solve your equation to find the value of z.

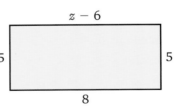

Learn... 6.2 Harder equations

Some equations have x terms on both sides.

Follow these steps to solve the equation.

- Collect together, on one side, all the terms that contain x.
- Collect together, on the other side, all the terms that do not contain x.
- Solve by doing the same to both sides of the equation, as in Learn 6.1.
- Check your answer by substituting it back into the equation.

Remember that a sign belongs to the term **after** it.

Take one step at a time. Do not try to do two steps at once.

Collect the terms in x on the side that has the largest number of them already.

Be extra careful if the equation involves negative amounts of the letter,
e.g. in an equation containing both $-4x$ and $2x$, you would collect the terms on the side where $2x$ is.

$2x$ is larger than $-4x$.

It helps to think of the number line. Which of the numbers is further to the right?

Example: Solve the equation $9 - 4y = -2y + 7$

Hint

$-2y$ is larger than $-4y$ so collect the y terms on the right-hand side.

Solution:

$$9 - 4y + 4y = -2y + 7 + 4y$$ Add $4y$ to both sides.
                       (This collects all the y terms on the right-hand side.)
$$9 = 2y + 7$$
$$9 - 7 = 2y + 7 - 7$$ Take 7 from both sides.
$$2 = 2y$$
$$1 = y$$
$$y = 1$$ Write the equation with y on the left-hand side.

Check

     LHS: $9 - 4 \times 1 = 5$
     RHS: $-2 \times 1 + 7 = 5$ left-hand side = right-hand side
     LHS = RHS ✓   They both have a value of 5.

Example: Find:

   a   the width

   b   the length of this rectangle.

All dimensions are in centimetres.

Solution:

   a   The opposite sides of a rectangle are equal in length.

     So $6b - 2.5 = 3.5 + b$ Collect the b terms on the left-hand side.
     $6b - 2.5 - b = 3.5 + b - b$ Take b from both sides.
     $5b - 2.5 = 3.5$
     $5b - 2.5 + 2.5 = 3.5 + 2.5$ Add 2.5 to both sides.
     $5b = 6$
     $\dfrac{5b}{5} = \dfrac{6}{5}$ Divide both sides by 5.
     $b = 1.2 \text{ cm}$ This is the width of the rectangle as shown on the diagram.

     The width of the rectangle is 1.2 cm.

   b   Length = $3.5 + b$ Substitute for b in one of the expressions for the length shown on the diagram.
              $= 3.5 + 1.2$
              $= 4.7$

     The length of the rectangle is 4.7 cm.

Hint

When finding the width, you could use the other expression, $6b - 2.5$

This is more complicated so choose the easier one.

The harder expression can be used as a check.

Check

$6b - 2.5 = 6 \times 1.2 - 2.5 = 7.2 - 2.5 = 4.7$ ✓

AQA *Examiner's tip*

It is good practice to do a check whenever you solve an equation.

Practise... 6.2 Harder equations

D

1 Solve these equations.

a $3x + 1 = x + 13$ e $2 + 2p = 4p - 1$

b $6y + 4 = -24 - y$ f $5b + 16 = 8b + 10$

c $4z + 1.5 = 2z - 3$ g $-7c - 3 = 30 - 4c$

d $8t - \frac{1}{2} = 4t + \frac{1}{2}$ h $10d - 0.6 = 0.9 - 5d$

2 Jared solves the equation $9x + 7 = 9 - x$
His first step is $8x + 7 = 9$
What mistake has Jared made?

3 Helen solves the equation $3y - 4 = 6 + 2y$
She gets the answer $y = 2$
Can you find Helen's mistake?

4 Work out the value of x.

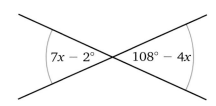

5 The perimeters of the equilateral triangle and the rectangle are equal.

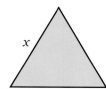
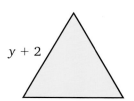

Work out the value of x.

6 The perimeters of the regular hexagon and the equilateral triangle are equal.

a Use this information to write down an equation in y.

b Solve your equation to find the value of y.

c What is the actual perimeter of each of the shapes?

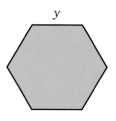

7 The perimeters of each of these shapes are equal.

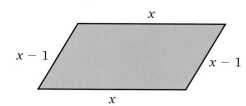

a Use this information to write down an equation in x.

b Solve your equation to find the value of x.

c What is the actual perimeter of each of the shapes?

8 Work out the value of x.

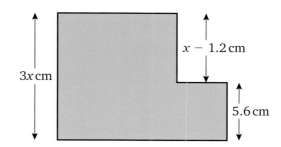

9 The diagram shows a rectangular lawn. The gardener has dug up a rectangular area. He is going to use this as a vegetable patch.

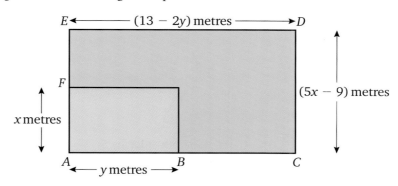

B and F are the midpoints of the sides AC and AE respectively.

a Form two equations, one in x and one in y.

b Solve your equations to find the values of x and y.

c What are the dimensions of the vegetable patch?

d What are the dimensions of the outer rectangle, the whole garden?

10 The line EF intersects the lines AB and CD.
The angles are as shown on the diagram.
Is AB parallel to CD?
Show working to justify your answer.

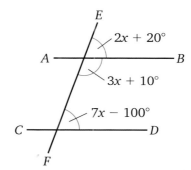

Learn... 6.3 Equations with brackets

Whenever you are asked to solve equations with **brackets**, you will usually begin by removing the brackets.

Usually they are removed by **expanding** the brackets.

×	x	+4
2	$2x$	+8

or $2(x + 4) = 2x + 8$

Multiply everything inside the bracket by the number outside the bracket.

For example, $2(x + 4)$ becomes $2x + 8$

and $-6(3x - 1)$ becomes $-18x + 6$.

Sometimes you can remove brackets by dividing the equation by a number. (See the alternative method in the first example overleaf.)

Example: Solve the equation $5(2x - 3) = 25$

Solution:

$$5(2x - 3) = 25$$ Expand the brackets first, then follow the rules for solving equations.

$$10x - 15 = 25$$ Remember to multiply both terms in the bracket by 5.

$$10x - 15 + 15 = 25 + 15$$ Add 15 to both sides.

$$10x = 40$$

$$\frac{10x}{10} = \frac{40}{10}$$ Divide both sides by 10.

$$x = 4$$

Check

$$5(2 \times 4 - 3) = 5(8 - 3) = 5 \times 5 = 25 \checkmark$$

Alternative method:

$$5(2x - 3) = 25$$

$$\frac{5(2x - 3)}{5} = \frac{25}{5}$$ Divide both sides by 5.

$$2x - 3 = 5$$

$$2x - 3 + 3 = 5 + 3$$ Add 3 to both sides.

$$2x = 8$$

$$\frac{2x}{2} = \frac{8}{2}$$ Divide both sides by 2.

$$x = 4$$

Check

$$5(2 \times 4 - 3) = 5(8 - 3) = 5 \times 5 = 25 \checkmark$$

This alternative method is only worth using here because 25 is divisible by 5.

Example: Solve the equation $2(y - 4) - 1 = 12 - 5y$

Solution:

$$2y - 8 - 1 = 12 - 5y$$ Expand the brackets first. $2 \times -4 = -8$

$$2y - 9 = 12 - 5y$$ Simplify any like terms.

$$2y - 9 + 9 = 12 - 5y + 9$$ Add 9 to both sides.

$$2y = 21 - 5y$$

$$2y + 5y = 21 - 5y + 5y$$ Add 5y to both sides.

$$7y = 21$$

$$\frac{7y}{7} = \frac{21}{7}$$ Divide both sides by 7.

$$y = 3$$

Check

$$\text{LHS: } 2(3 - 4) - 1 = 2 \times (-1) - 1 = -2 - 1 = -3$$

$$\text{RHS: } 12 - 5 \times 3 = 12 - 15 = -3$$

$$\text{LHS} = \text{RHS} \checkmark$$

Bump up your grade

Students working at Grade C should be able to solve equations which have brackets and the unknown appearing on both sides.

Example: This shape is made up of two rectangles A and B. The total area of the shape is $18\,cm^2$.

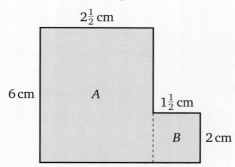

$(x + 1)$ cm

6 cm A

x cm

B 2 cm

a Write down an equation for the area of the shape in terms of x.

b Solve your equation to find the value of x.

c Redraw your shape replacing all measurements with numbers.

This does not have to be to scale.

Solution:

a The area consists of two rectangles.
Begin by writing down the area for each of them separately.

Area of $A = 6(x + 1)$ Area of $B = 2x$

Area of A + Area of B = Total area

$$6(x + 1) + 2x = 18$$

b $6x + 6 + 2x = 18$ Expand the brackets.

$8x + 6 = 18$ Simplify by collecting x terms.

$8x + 6 - 6 = 18 - 6$ Take 6 from both sides.

$8x = 12$

$\dfrac{8x}{8} = \dfrac{12}{8}$ Divide both sides by 8.

$x = 1\frac{4}{8}$ or $1\frac{1}{2}$

c Width of $A = x + 1 = 2\frac{1}{2}$ cm

Width of $B = x = 1\frac{1}{2}$ cm

$2\frac{1}{2}$ cm

6 cm A

$1\frac{1}{2}$ cm

B 2 cm

AQA *Examiner's tip*

Don't try to do two steps at once. Most students make mistakes if they rush their working.

Bump up your grade

For a Grade C you must be able to solve equations involving fractions and decimals.

Practise... 6.3 Equations with brackets

G F E D C

1 Solve these equations.

a $4(2x + 1) = 44$

b $2(y - 3) = 32$

c $35 = 7(3z - 1)$

d $-15 = 3(6a + 4)$

e $11(3b + 1) = 44$

f $45 = 5(c + 1.5)$

2 Solve these equations.

a $8b + 7 = 3(3b - 1)$

b $12 + 3c = 8(c - 1)$

c $3 - 2d = 4(2d - 7)$

d $0.1f - 6.5 = 2(0.8f - 1)$

e $4(3y - 2) = 3(3y + 1)$

f $8(j - 1) = 5(j - 2)$

g $5(t + 1) = 3(t - 3)$

D

C

C

3 Emma thinks of a number, x, adds 5 and then doubles the result.

Her answer is 40.

Write down and solve an equation in x to work out Emma's number.

4 Dan thinks of a number, x, subtracts 2 and then multiplies the result by 5.

His answer is 30.

Write down and solve an equation in x to work out Dan's number.

5 These rectangles have the same area.

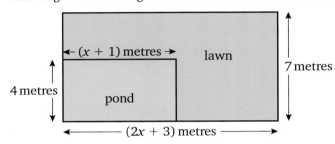

Work out the value of x.

⚠ 6 The diagram shows a garden. A section of the lawn has been removed to make a pond.

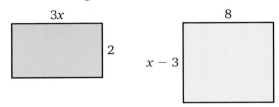

The area of the lawn is 35 m².

a Write down an equation for the area of the lawn.

b Solve your equation to find the value of x.

c Find the dimensions of the pond and the whole garden.

d How can you check this? Were you correct?

⚠ 7 The area of this triangle is 64 cm².

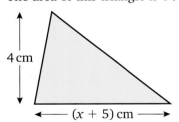

a Write down an equation, in x, for the area of the triangle.

b Solve your equation to find the value of x.

c Find the length of the base of the triangle.

 8 The following shape is made from equilateral triangles of sides $(x - 2)$ cm.

The perimeter is 42 cm.

a Write down an equation in x.

b Work out the value of x.

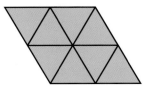

Learn... 6.4 Equations with fractions

Fractions are removed by multiplying both sides by the **denominator**.

For example, if the equation contains $\frac{x}{4}$, you would multiply by 4.

If there is more than one fraction, say $\frac{3x}{5}$ and $\frac{x}{2}$, you would multiply everything by 5 throughout and then by 2.

An alternative method would be to multiply, just the once, by the **lowest common denominator**.

The lowest common denominator for these two fractions is 10. This is because 10 is the smallest number which both 5 and 2 divide into exactly.

Harder equations have numerators with more than one term on the top of the fraction, e.g. $\frac{3x - 2}{4}$

There are 'invisible brackets' around the terms on top of an algebraic fraction.

To make this clear, you should put them in: $\frac{(3x - 2)}{4}$. Then you are less likely to make a mistake.

Example: Solve the equation $\frac{x}{4} + 3 = 5$

Solution: This is an example of the simplest type of equation with a fraction.

With only one fraction in the equation, you should first get the fraction on its own on one side of the equation and then multiply by 4.

You are working towards finding $x = ...$, so $+ 3$ is the first term to go from the left-hand side of the equation.

$\frac{x}{4} + 3 - 3 = 5 - 3$ Take 3 from both sides.

$\frac{x}{4} = 2$ Now the fraction term is on its own.

$\frac{x}{4} \times 4 = 2 \times 4$ Multiply both sides by 4 (the denominator).

$x = 8$

Check

$\text{LHS} = \frac{8}{4} + 3 = 2 + 3 = 5 = \text{RHS} ✓$

> **Hint**
> To remove the denominator of the fraction, multiply both sides by 4.

Example: Solve the equation $\frac{3x - 2}{2} = 5$

Solution: This is an example of an equation with more than one term on the top of the fraction.

The whole of $3x - 2$ is divided by 2.

$\frac{(3x - 2)}{2}$ is the same as one half of $3x - 2$.

$2 \times \frac{(3x - 2)}{2} = 2 \times 5$ Multiply both sides by 2.

$3x - 2 = 10$

$3x - 2 + 2 = 10 + 2$ Add 2 to both sides.

$3x = 12$

$\frac{3x}{3} = \frac{12}{3}$ Divide both sides by 3.

$x = 4$

Check

$\text{LHS} = \frac{(3 \times 4 - 2)}{2} = \frac{(12 - 2)}{2} = \frac{10}{2} = 5 = \text{RHS} ✓$

> **AQA Examiner's tip**
> You can put in the invisible brackets before you start your working. Here the brackets have been put round $3x - 2$.

Example: A stick is $2x + 15$ cm long.

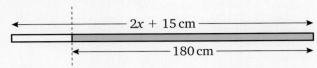

If one fifth of the stick is cut off, the stick will now be 180 cm long.

How long was the stick before the piece was cut off?

Solution: One fifth of the stick was cut off so four fifths remains. This remaining part is 180 cm.

This can be turned into an equation.

$$\frac{4(2x + 15)}{5} = 180$$

$$\frac{(8x + 60)}{5} = 180 \qquad \text{Multiply out the brackets.}$$

$$\frac{5 \times (8x + 60)}{5} = 5 \times 180 \qquad \text{Multiply both sides by 5.}$$

$$8x + 60 = 900$$

$$8x + 60 - 60 = 900 - 60 \qquad \text{Take 60 from both sides.}$$

$$8x = 840$$

$$\frac{8x}{8} = \frac{840}{8} \qquad \text{Divide both sides by 8.}$$

$$x = 105 \text{ cm}$$

This means that the stick was originally $2 \times 105 + 15 = 210 + 15 = 225$ cm long.

Check

$$\text{LHS} = \frac{4(2 \times 105 + 15)}{5} = \frac{4(225)}{5} = 4 \times 45 = 180 = \text{RHS} \checkmark$$

Practise...   **6.4 Equations with fractions** (k!)   G F E D C

C

1   Solve these equations.

a $\frac{x}{3} - 1 = 5$ **d** $\frac{5x + 1}{3} = 12$ **g** $\frac{x}{4} + \frac{x}{3} = 7$

b $\frac{y}{9} + 4 = 9$ **e** $\frac{2y - 3}{4} = 6$ **h** $\frac{y}{3} - \frac{y}{5} = 4$

c $\frac{f}{2} + 5 = -3$ **f** $7 = \frac{11 - z}{5}$ **i** $\frac{5z}{9} + \frac{z}{3} = 8$

2

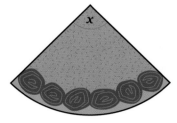

Ravi decides to share his piece of cake with his friends.

The original piece made an angle of x at the centre.

He took one-third of this slice. The angle at the centre of his slice was 25°.

a   Use this information to write down an equation in x.

b   Solve this equation to find the angle at the centre of Ravi's original piece of cake.

3 David has a large piece of cake.

He helps himself to one-quarter of it. Once he has taken this, the angle at the centre is reduced to 207°.

Work out the value of x.

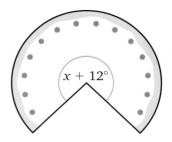

4 A farmer is building somewhere for his prize pig to live.

The area consists of a pig sty for the pig to shelter in, and a run.

All measurements are in metres.

One-quarter of the pig's accommodation is taken up by the pig sty.

The run has an area of $13.5\,m^2$.

a Write down an expression for the total area.

b Write down an expression for the area of the run.

c Using the information given, form an equation in x and solve it to find the value of x.

d Find the total area of the pig sty and the run.

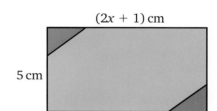

5 The red areas make up one-sixth of the total area of the rectangle.

Each of the red parts has an area of $7\,cm^2$.

Work out the perimeter of the rectangle.

6 A factory is making metal brackets.

It cuts them from a sheet of material $2x\,cm$ by $3x\,cm$.

The remaining material, shown in purple, is the metal that is wasted.

The factory wastes one-third of the material every time they cut out a bracket.

The area of the wasted material is $450\,cm^2$.

a Use the information to write down an equation for the wasted material.

b Work out the dimensions of the metal sheet.

7 There are two bags containing sweets.

One contains $x - 1$ sweets and the other contains $2x + 4$ sweets.

Hannah takes half of the sweets in the first bag and a third of the sweets in the second bag.

She takes nine sweets altogether.

a Find the value of x.

b Find out how many sweets there were in each bag to start with.

$x - 1$ sweets $2x + 4$ sweets

6 Assess (k!)

F

1 Solve these equations.

   a $3x = 24$ **b** $y - 8 = 3$ **c** $2z = 5$

E

2 Solve these equations.

   a $\dfrac{a}{5} = 2$ **c** $6c - 5 = 13$ **e** $4 = 7 - 2e$

   b $\dfrac{b}{6} = -5$ **d** $3 + 2d = 1$

D

3 Solve these equations.

   a $7q - 2 = 4q + 7$ **d** $5(p + 3) = 35$

   b $9m + 7 = 4m - 3$ **e** $24 = 3(2t - 3)$

   c $4n - 9 = 2 - 7n$ **f** $4(2v + 1) = 3(5v - 8)$

4 The diagram contains two parallel lines.

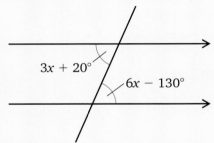

$3x + 20°$

$6x - 130°$

Write down and solve an equation in x.

C

5 Solve these equations.

   a $4(x + 3) + 3(2x - 1) = 39$ **d** $3 - \dfrac{z}{2} = 7$

   b $5(2y - 1) = 1 + 2(y + 3)$ **e** $\dfrac{5d - 4}{3} = 1$

   c $\dfrac{t}{6} + 3 = 8$

6 The diagram shows a rectangular garden.

There is a rectangular vegetable patch in the garden.
The rest of the garden is lawn.

The area of the lawn is 9.5 m².

Work out the area of the vegetable patch.

$(3x + 1)$ metres

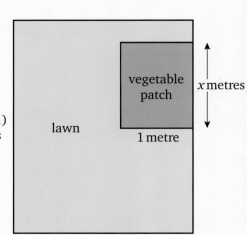

vegetable patch

x metres

lawn

1 metre

2 metres

AQA Examination-style questions

1 The total of each row is given at the side of the table.

$4x + 1$	$2(x + 5)$	20
$2x$	4	A

Find the values of x and A. *(3 marks)*

AQA 2007

2 Dean picks three numbers, which total 77.
His first number is y. His second number is five more than his first number. His third number is double his first number.
Work out his three numbers. *(3 marks)*

AQA 2007

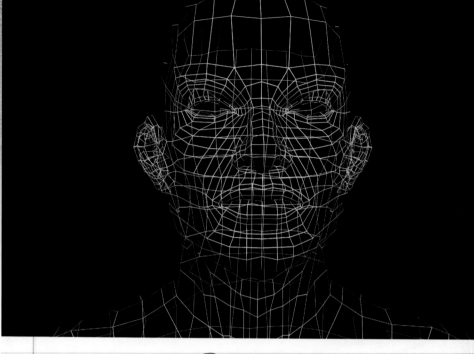

Objectives

Examiners would normally expect students who get these grades to be able to:

G

recognise and name shapes such as parallelogram, rhombus, trapezium

E

calculate interior and exterior angles of a quadrilateral

D

classify a quadrilateral using geometric properties

C

calculate exterior and interior angles of a regular polygon.

Did you know?

Polygons and video games

Objects in video games are made up of lots of polygons. Pictures are made up of a series of polygons such as triangles, squares, rectangles, parallelograms and rhombuses. The more polygons there are, then the better the picture looks.

The polygons are all given coordinates. The computer changes and rotates the coordinates to match your position in the game. This gives the impression of movement.

For example, if you are far away, the computer shrinks all the coordinates of the polygons. This makes the polygons appear smaller on the screen so they look further away.

Key terms

quadrilateral
polygon
exterior angle
interior angle
diagonal
bisect
perpendicular
pentagon
hexagon
regular
octagon
decagon
nonagon

You should already know:

✔ how to use properties of angles at a point, angles on a straight line, perpendicular lines, and opposite angles at a vertex

✔ the differences between acute, obtuse, reflex and right angles

✔ how to use parallel lines, alternate angles and corresponding angles

✔ how to prove that the angle sum of a triangle is 180°

✔ how to prove that the exterior angle of a triangle is equal to the sum of the interior opposite angles

✔ angle properties of equilateral, isosceles and right-angled triangles.

 **Learn...   7.1 Properties of quadrilaterals**

A **quadrilateral** is a **polygon** with four sides.

You need to know the names and properties of the following special quadrilaterals.

Square – a quadrilateral with four equal sides and four right angles

Trapezium – a quadrilateral with one pair of parallel sides

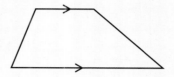

Rectangle – a quadrilateral with four right angles, and opposite sides equal in length

Parallelogram – a quadrilateral with opposite sides equal and parallel

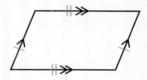

Kite – a quadrilateral with two pairs of equal adjacent sides

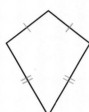

Rhombus – a quadrilateral with four equal sides and opposite sides parallel

Isosceles trapezium – a trapezium where the non-parallel sides are equal in length

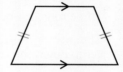

All quadrilaterals have four sides and four angles.

A quadrilateral can be split into two triangles.

The angles in a triangle add up to 180°.

The quadrilateral is made up of two triangles.

The angles in a quadrilateral add up to 2 × 180° = 360°

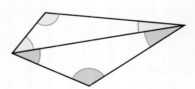

Example: Calculate the angles marked with letters in this shape.

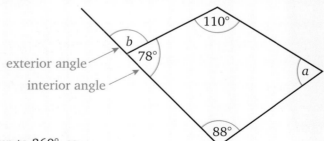

exterior angle

interior angle

Solution: The angles in the quadrilateral add up to 360°, so

$a = 360° - (78° + 88° + 110°)$

$\quad = 360° - 276°$

$\quad = 84°$

The **exterior** and **interior angles** add up to 180°, so

$b = 180° - 78°$

$\quad = 102°$

AQA Examiner's tip

Always make sure that your answer is reasonable. Angle b is an obtuse angle so that answer is reasonable.

Practise...

7.1 Angle properties of quadrilaterals

G

1 Write down the mathematical name of these quadrilaterals.

a

b

c

d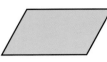

e

f

F

2 Small metal rods can be joined at the ends to make shapes.
The rods are all the same length.
Three rods can be used to make an equilateral triangle like this.

Alan uses four rods.
Write down the names of the shapes that he can make.

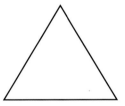

E

3 Calculate the angles marked with letters.

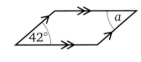

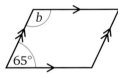

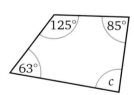

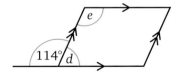

Not drawn accurately

4 a Three angles of a quadrilateral are 60°, 65° and 113°.
Work out the size of the fourth angle.

b Two angles of a quadrilateral are 74° and 116° and the other two angles are equal.
Work out the size of the other two angles.

c i If all four angles of a quadrilateral are equal, what size are they?

ii What sorts of quadrilateral have four equal angles?

5   Calculate the angles *a*, *b*, *c*, *d* and *e* in these quadrilaterals.

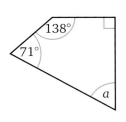

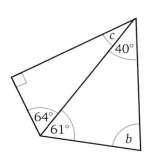

Not drawn
accurately

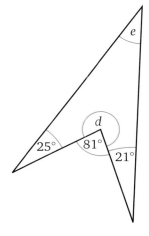

6   square   rectangle   parallelogram   rhombus   kite   trapezium

a   Which of these quadrilaterals have all four sides equal?

b   Which of these quadrilaterals have opposite sides which are parallel?

c   Which of these quadrilaterals have adjacent sides which are equal?

d   Which of these quadrilaterals has only one pair of parallel sides?

7   Barry measures the angles of a quadrilateral. He says that three of the angles are 82° and the other one is 124°. Is he right?
Give a reason for your answer.

8   Harry measures the angles of a quadrilateral. He says that the angles are 72°, 66°, 114° and 108°. He says the shape is a trapezium. Is he right?
Give a reason for your answer.

9   A cyclic quadrilateral is a quadrilateral where all four vertices (corners) can be drawn on the circumference of a circle.

Which of the following are cyclic quadrilaterals?

square   rectangle   parallelogram   rhombus   kite

10   Tracey says that it is possible for a trapezium to be a cyclic quadrilateral.

Is she correct?

Give a reason for your answer.

11   One angle of a quadrilateral is 150° and the other three angles are equal.

Write down and solve an equation in *x*.

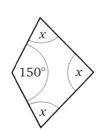

12   *EDC* is a straight line and angle *DAB* = angle *ABC*

Work out angle *ABC*.

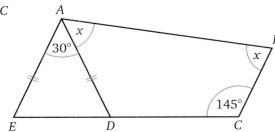

13   Charlie has two pieces of card in the shape of equilateral triangles. Each side is 4 cm long.

He cuts each piece of card in half along the dotted lines as shown.

He now has four right-angled triangles.

Each triangle is exactly the same.

   a   Use the four triangles to make:

      i   a rectangle   **iii**   a parallelogram

      ii   a trapezium   **iv**   a rhombus.

     Draw a diagram to show each of your answers.

   b   Work out the size of the angles in each of the shapes you made in part **a**.

14   **a**   A kite always has an obtuse angle. True or false?
      Give a reason for your answer.

   b   Can a kite have two obtuse angles?
      Give a reason for your answers.

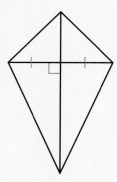

Learn...   7.2   Diagonal properties of quadrilaterals

A **diagonal** is a line joining one vertex (corner) of a quadrilateral to another.

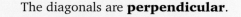

    Each quadrilateral has two diagonals.

    The square has two diagonals.

    The diagonals are the same length.

    The diagonals **bisect** one another.   Bisect means they cut one another in half.

    The diagonals are **perpendicular**.   Perpendicular means at right angles.

Example: What is the mathematical name of this quadrilateral?

   •   The diagonals are different lengths.

   •   The diagonals are at right angles to each other.

   •   Only one diagonal is bisected by the other.

Solution:

Drawing the diagonals using the information given, you can see what the shape is.

The quadrilateral is a kite.

Practise...

7.2 Diagonal properties of quadrilaterals

k!

G F E D C

1   **a**   Copy each of these shapes and draw their diagonals.

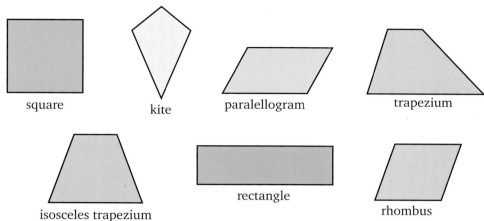

square   kite   paralellogram   trapezium

isosceles trapezium   rectangle   rhombus

b   Copy and complete this table.

Shape	Are the diagonals equal? (Yes/No)	Do the diagonals bisect each other? (Yes/No/Sometimes)	Do the diagonals cross at right angles? (Yes/No)	Do the diagonals bisect the angles of the quadrilateral? (Yes/No/Sometimes)
Square				
Kite				
Parallelogram				
Trapezium				
Isosceles trapezium				
Rectangle				
Rhombus				

2   Rajesh has drawn a quadrilateral. Its diagonals are equal.

What shapes might he have drawn?
(Use the table from Question 1 to help you.)

3   Michelle says that the diagonals of a rectangle bisect the angles.

So angles a and c are both 45°, and angle b must be 90°.

Is she right? Give a reason for your answer.

Not drawn accurately

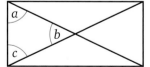

D
C

4   The diagram shows a rhombus $ABCD$. AC and BD are the diagonals. Angle $ADB = 32°$

Calculate angle DAC.

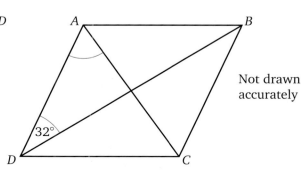

Not drawn accurately

E

D

C

5 square   rectangle   parallelogram   rhombus   kite   trapezium

a Which of these quadrilaterals have diagonals of different lengths?

b Which of these quadrilaterals have diagonals that cross at right angles?

c Which of these quadrilaterals have all four sides equal and diagonals that bisect at right angles?

d Which of these quadrilaterals have opposite sides which are parallel and diagonals of different lengths?

6 Calculate the angles *a–f* in the diagrams.

Give a reason for each answer.

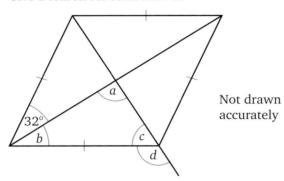

Not drawn accurately

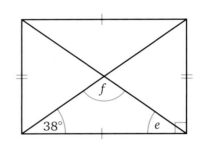

7 Copy and complete this table. The first row has been done for you.

Shape	Number of different length sides (at most)	Number of right angles (at least)	Pairs of opposite sides parallel	Diagonals must be equal	Diagonals bisect each other	Diagonals cross at right angles
Square	1	4	Both	Yes	Yes	Yes
Rectangle						
Trapezium						
Rhombus						
Parallelogram						
Kite						
Isosceles trapezium						

 Learn...   **7.3 Angle properties of polygons** *k!*

The interior angles of a triangle add up to 180°.

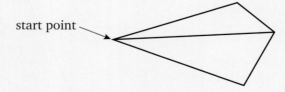

start point

A quadrilateral has four sides and can be split into two triangles by drawing diagonals from a point.

The sum of the angles is 2 × 180° = 360°

A **pentagon** has five sides and can be split into three triangles by drawing diagonals from a point.

The sum of the angles is $3 \times 180° = 540°$

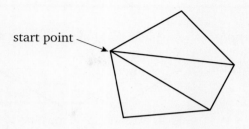

start point

A **hexagon** has six sides and can be split into four triangles by drawing diagonals from a point.

The sum of the angles is $4 \times 180° = 720°$

In general a polygon with n sides can be split into $(n - 2)$ triangles.

The sum of the angles is $(n - 2) \times 180°$.

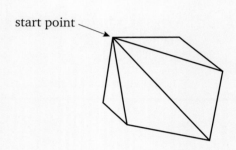

start point

The **interior angles** of a polygon are the angles inside the polygon.

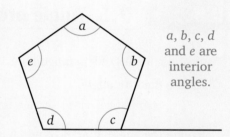

a, b, c, d and e are interior angles.

The **exterior angles** of a polygon are the angles between one side and the extension of the side.

The exterior angles of a polygon add up to $360°$.

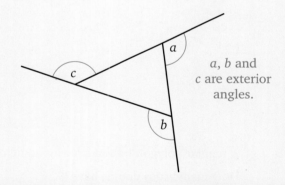

a, b and c are exterior angles.

Example: Find the interior angle of a **regular octagon**.

Solution: **Either:**

An octagon has eight sides.

So the sum of the angles is $(8 - 2) \times 180° = 1080°$

A regular octagon has all angles equal, so each angle is $1080° \div 8 = 135°$

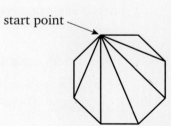

start point

Or:

A regular octagon has eight equal exterior angles.

So each exterior angle is $360° \div 8 = 45°$

So each interior angle is $180° - 45° = 135°$

Example: A regular polygon has interior angles of 144°. How many sides does it have?

Solution:

interior angle

exterior angle

144°

Each exterior angle must be 180° − 144° = 36°

The exterior angles of a convex polygon add up to 360°.

A regular polygon has all sides equal and all angles equal.

So there must be 360° ÷ 36° = 10 exterior angles

The polygon has 10 sides.

AQA *Examiner's tip*

Always draw a diagram to help answer the questions.

You can then label the diagram to keep track of what you know.

Practise... 7.3 Angle properties of polygons 🔑   G F E D C

C

1   Four of the angles of a pentagon are 110°, 130°, 102° and 97°.

Calculate the fifth angle.

2   Calculate the angles marked *a* and *b* in the diagram.

Explain how you worked them out.

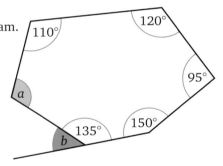

110°   120°

95°   Not drawn accurately

a

135°   150°

b

3   A regular polygon has an exterior angle of 60°.

How many sides does it have?

4   Calculate the difference between the interior angle of a regular **decagon** (ten-sided shape) and the interior angle of a regular **nonagon** (nine-sided shape).

5   James divides a regular hexagon into six triangles as shown.

He says the angle sum of a regular hexagon is 6 × 180°.

Is he correct?

Give a reason for your answer.

6   Lisa says that a regular octagon can be split into two trapeziums and a rectangle as shown.

She says the angle sum of the octagon is 3 × 360°.

Show that Lisa is correct.

7 The diagrams show how you draw an equilateral triangle and a regular pentagon inside a circle. You do this by dividing the angle at the centre equally.

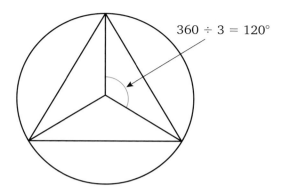

$360 \div 3 = 120°$

$360 \div 5 = 72°$

Use the same method to draw a regular hexagon and a regular nonagon (nine-sided shape) inside a circle.

8 The diagram shows a regular pentagon *ABCDE* and a regular hexagon *DEFGHI*.

Calculate:

a angle *EDC*

b angle *EDI*

c obtuse angle *CDI*

d angle *BAC*

e angle *CAE*

f angle *HIG*

g angle *DIG*.

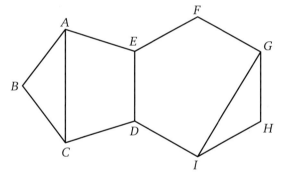

9 A badge is in the shape of a regular pentagon.

The letter V is written on the badge.

What is the size of the angle marked *x*?

10 A company makes containers as shown.

The top is in the shape of a regular octagon.

a What is the size of each interior angle?

b When the company packs them into a box, will they tessellate (fit together exactly)? If not, what shape will be left between them?

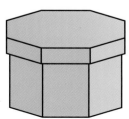

7

Assess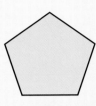

G

1 Write down the mathematical name of each of these shapes.

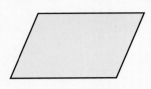

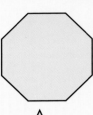

F

2 Write down the letters of the shapes in Question 1 that have:

a   some sides equal (but not all)

b   all sides equal

c   any acute angles

d   any obtuse angles

e   some equal angles

f   any adjacent sides equal

g   all diagonals equal

h   diagonals perpendicular to each other

i   diagonals of different lengths

j   any adjacent angles equal.

E

3 Calculate the angles marked a, b and c in this parallelogram.

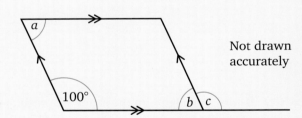

Not drawn accurately

D

4 Which of the following polygons are possible and which ones are not possible?

Make an accurate drawing of each one that is possible.

a   A kite with a right angle

b   A kite with two right angles

c   A trapezium with two right angles

d   A trapezium with only one right angle

e   A triangle with a right angle

f   A triangle with two right angles

g   A pentagon with one right angle

h   A pentagon with two right angles

i   A pentagon with three right angles

j   A pentagon with four right angles.

D
C

5 Find the value of the angles marked in these diagrams.

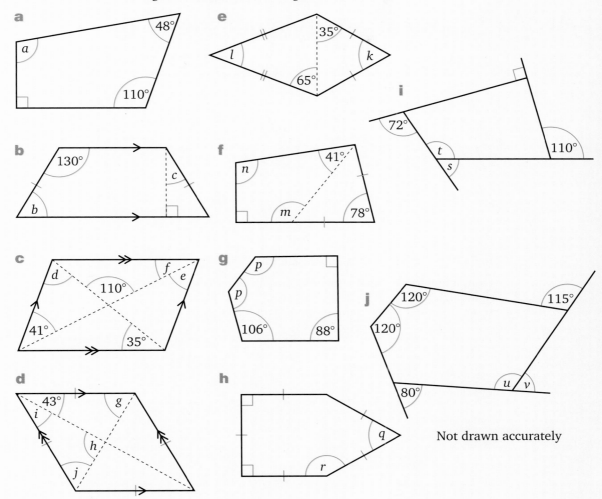

Not drawn accurately

6 Sophie says her regular polygon has an exterior angle of 40°.

Adam says that is not possible.

Who is correct?

Give a reason for your answer.

C

7 The exterior angle of a regular polygon is 4°.

a How many sides does the polygon have?

b What is the size of each interior angle in the polygon?

c What is the sum of the interior angles of the polygon?

8 *ABCDE* is a regular pentagon.

DEG, *DCF* and *GABF* are straight lines.

Work out the size of angle *x*.

Not drawn accurately

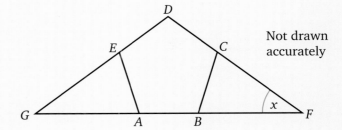

AQA Examination-style questions 🔑

1   a   The diagrams show the diagonals of two different quadrilaterals.

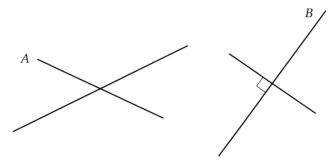

Write down the names of these quadrilaterals.   *(3 marks)*

   b   i   Draw a quadrilateral that has only one pair of parallel lines and exactly two right angles.   *(2 marks)*

        ii   Write down the name of this quadrilateral.   *(1 mark)*

AQA 2004

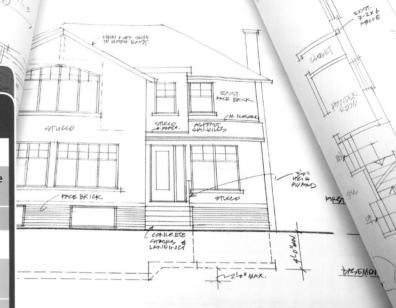

Objectives

Examiners would normally expect students who get these grades to be able to:

G

use coordinates in the first quadrant, such as plotting the point (2, 1)

recognise the net of a simple solid

F

use coordinates in all four quadrants, such as (2, −1), (−2, −3) and (−2, 1)

use simple real-life graphs, such as read values from conversion graphs

draw the net of a simple shape, such as a matchbox tray

E

draw lines such as $x = 3$, and $y = x$

interpret horizontal lines on a distance–time graph

use real-life graphs to find values, such as find distances from distance–time graphs

draw a simple shape, such as a cuboid, on isometric paper

D

make simple interpretations of real-life graphs

draw the front elevation, side elevation and plan of a solid on squared paper

C

carry out further interpretation of real-life graphs, for example find the average speed in km/h from a distance–time graph over time in minutes.

Did you know?

Draw the right house, build the house right

Plans and elevations are very important to people who design and build houses.

When a house is built for someone, an architect will draw out accurate plans to ensure they get the house they want, and so they can see what it will look like when it is finished.

The builder uses the plans to build the house. Following the plans carefully will ensure that the walls, doors, and windows are in the correct place.

Key terms

coordinates
origin
axis
vertex
edge
face
plan
elevation
net

You should already know:

✔ negative numbers and the number line

✔ the meaning of the words vertex, vertices, axis, axes, horizontal and vertical

✔ names of common quadrilaterals and their properties

✔ how to substitute into a formula

✔ 3-D shapes such as cube, cuboid, tetrahedron, pyramid and prisms.

Learn... 8.1 Coordinates and equations of a straight line

Coordinates are used to describe the position of a point relative to a starting point called the **origin** labelled *O* in the diagram.

The horizontal **axis** is called the *x*-axis, coloured blue in the diagram.

The vertical axis is called the *y*-axis, coloured red in the diagram.

The axes divide the graph paper into four quadrants (quarters).

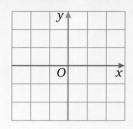

The next diagram shows the first quadrant, where *x*- and *y*-values are both positive.

The coordinates of point *B* are (4, 3). The *x*-coordinate is always first. They are given in alphabetical order, *x* before *y*. The coordinates are always separated by a comma, and have brackets around them to keep them together as a pair.

The coordinates of *C* and *A* are (3.5, 2) and (1.5, 1.5). It is acceptable to use either decimals or fractions when the coordinates are not integers.

Take care not to confuse *D* which is (4, 0) and *E* which is (0, 4).

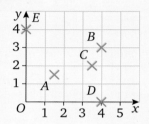

AQA *Examiner's tip*

> Sometimes students forget which axis is *x* and which is *y*. To help you remember which is which you need to 'Wise up, X is a cross' (*y*'s up, *x* is across).

This diagram shows all four quadrants.
Some points have negative coordinates.

Point *A* is in the first quadrant (as above) and has coordinates (1, 4).

Point *B* has coordinates (−1, 4).

Point *C* is (−3, −1).

Point *D* is (1, −3).

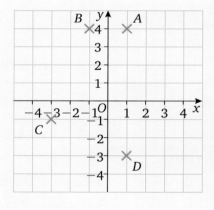

Equations of horizontal and vertical lines

The equation of a straight line is a 'rule' that applies to every point on that line.

The coordinates of *A*, *B* and *C* are (1, 3), (2, 3) and (4, 3) respectively. They all have the *y*-coordinate = 3.
If you draw a line through them, $y = 3$ for every point on the line.
The equation for this line is $y = 3$.
This line is green on the diagram and is labelled $y = 3$

The coordinates of points *C*, *D* and *E* are (4, 3), (4, 0) and (4, 2). They all have the *x*-coordinate = 4.
If you draw a line through them, $x = 4$ for every point on the line.
The equation for this line is $x = 4$.
This line is blue on the diagram and is labelled $x = 4$

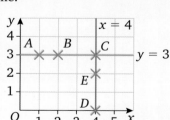

AQA *Examiner's tip*

> Always label the *x*- and *y*-axes. When you draw a straight line always remember to label it with its equation.

8.1 Coordinates and equations of a straight line

Practise...

G F E D C

1 Write down the coordinates of each of the points on the grid.

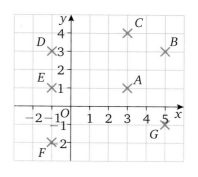

2 Draw a pair of x- and y-axes from -5 to 5 on squared paper.
Plot and label the following points.

$A(2, 3), B(4, 3), C(2, -1), D(-1, 2), E(-2, 2), F(2, -2), G(-2, -2)$

3 The diagram shows three vertices of a rectangle $ABCD$.
Write down the coordinates of D, the fourth
vertex of the rectangle?

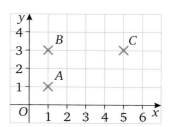

4 The diagram shows part of a pattern made of identical
rectangles.

a Write down the coordinates of the four vertices
of rectangle 3.

b Work out the coordinates of the four vertices
of rectangle 4.

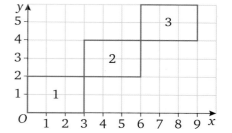

5 **a** Write down the equations of the lines on the grid.

b Declan said that line A crosses line D at the point $(1, 4)$.
What mistake has Declan made?

c Write down the coordinates of the points where:

i lines A and B cross

ii lines B and C cross

iii lines A and E cross

iv lines B and F cross

v lines D and F cross.

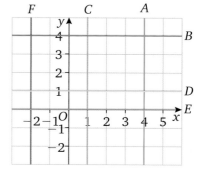

6 **a** Draw a pair of x- and y-axes from -5 to 5 on squared paper. Draw and label the lines with the
following equations.

i $y = 3$

ii $x = 2$

iii $y = -2$

iv $x = -1$

b Write down the coordinates of the points where the lines cross. What do you notice about the
coordinates of the points and the equations of the lines?

G
F

E

E
D

7 The equation $y = x$ tells you that the x- and y-coordinates are equal.
The point (3, 3) is on this line.

   a Give two other points that are on the line.

   b Draw a pair of x- and y-axes from -5 to 5 on squared paper.
Plot your points, join them and label the line.

   c What are the coordinates of the point where the line crosses the x-axis?

8 The equation $y = -x$ tells you that the x- and y-coordinates are the same number, but with different signs. For example $y = -2$ when $x = 2$. So $(2, -2)$ is on the line.

   a Write down the coordinates of two other points on this line.

   b Draw a pair of x- and y-axes from -5 to 5 on squared paper.
Plot your points, join them and label your line.

   c What do you notice about the lines you have drawn in this question and Question 7?

9 The diagram shows three of the vertices of a square.
What are the coordinates of the fourth vertex?

> **Hint**
> You may find it helpful to copy the diagram and draw the square.

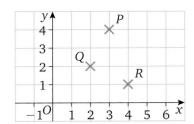

⚠ 10 The diagram shows three of the vertices of a parallelogram.

Write down the coordinates of the fourth vertex.
There is more than one possible place to put the fourth vertex. Find them all.

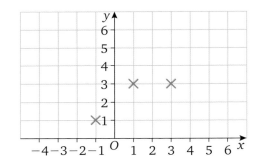

Learn...   8.2 Real-life graphs

Tables and formulae are often used to show connections between data. It is often easier to see at a glance what is going on when you use a graph.

Graphs can be used to represent many different situations.

Conversion graphs can be used to convert from one unit to another. This includes distances such as miles and kilometres, or currency conversions such as pounds (£) and euros (€).

The gradient of a straight line is found using the formula:

$$\text{gradient} = \frac{\text{increase in } y}{\text{increase in } x}$$

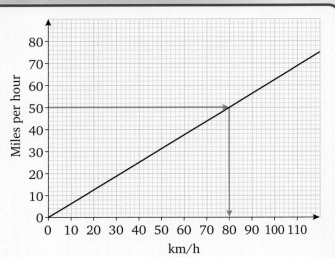

Distance–time graphs can be used to help calculate speeds. The gradient of a distance–time graph tells you the object's speed.

You can use this kind of graph see how far an object travelled in each unit of time.

You need to learn the formula:

$$\text{average speed} = \frac{\text{total distance}}{\text{total time}}$$

Graphs which show costs can be used to compare prices, such as the cost of different contracts for mobile phones. If cost is on the vertical axis and time in minutes on the horizontal axis, the gradient tells you the cost per minute.

Example: An oven is set at a temperature of 200°C. This graph shows the temperature inside the oven over a two-hour period.

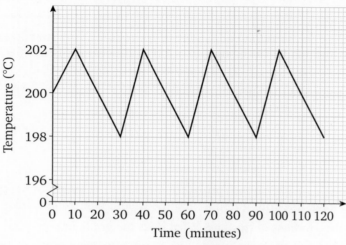

a   Explain what is happening in the graph.

b   What is the temperature in the oven after 45 minutes?

c   Between 30 and 40 minutes the oven is warming up.
    What does the gradient of this section tell you?

d   What does the gradient of the line between 70 minutes and 90 minutes tell you?

Solution: a   The graph tells you that the oven warms up, then cools down. When the graph goes up, this shows the temperature is rising. When the graph goes down this tells you the oven is cooling down. The temperature goes up more quickly than it cools down. You know this because the graph is steeper when the oven is warming up.

b

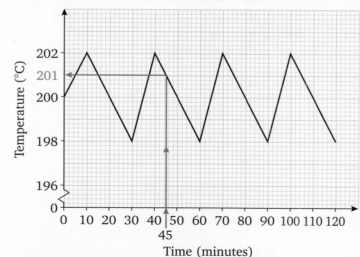

Read up from 45 minutes on the horizontal axis to the graph. From there read across to the vertical axis and write the temperature down. It is 201°C. This is shown with an arrow on the graph.

c Between 30 and 40 minutes the temperature rises from 198°C to 202°C.
The temperature rises 4°C in 10 minutes.

$$\text{The gradient} = \frac{\text{increase in } y}{\text{increase in } x} = \frac{4}{10} = 0.4$$

This tells you the temperature increases at a rate of 0.4°C per minute.
The gradient tells you the rate at which the temperature increases.

d The gradient of the line between 70 and 90 minutes tells you the rate at which the oven cools down.

$$\text{The gradient} = \frac{\text{increase in } y}{\text{increase in } x}$$

The temperature is decreasing, that is a negative increase. The temperature decreases by 4°C in 20 minutes.

$$\text{Gradient} = \frac{-4}{20} = -0.2 \text{ so the temperature is decreasing at 0.2°C per minute.}$$

Example: Benny is doing a science experiment looking at springs. A spring stretches when weights are added to it. The formula for its length is $L = 3M + 12$ where M is the mass in kg and L is the length in cm.

a Copy and complete the table.

M	2	4	6	8	10
L	18			36	

b Draw the graph for the values in the table.

c Benny hangs a weight on the spring. He measures the length of the spring.
It is 35 cm long. Use your graph to find the weight Benny has used.

d Calculate the gradient of this line. What does the gradient represent?

e The line meets the vertical y-axis at 12 cm. What does this tell you about the spring?

Solution: **a** Use the formula $L = 3M + 12$

For a 4 kg mass: $L = 3 \times 4 + 12 = 12 + 12 = 24$

For a 6 kg mass: $L = 3 \times 6 + 12 = 18 + 12 = 30$

For a 10 kg mass: $L = 3 \times 10 + 12 = 30 + 12 = 42$

M	2	4	6	8	10
L	18	24	30	36	42

b Plot the points carefully.

c Read across from 35 cm on the vertical axis to the graph.
From the graph read down to the horizontal axis. Write down the value of 7.6 kg.
The red arrows on the graph show this.

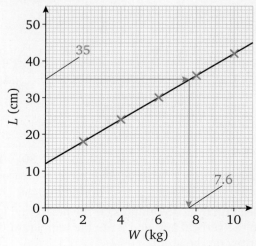

d   Using the formula, gradient $= \dfrac{\text{increase in } y}{\text{increase in } x}$

$$= \dfrac{(42 - 18)}{(10 - 2)}$$

$$= \dfrac{24}{8}$$

Gradient $= 3$

The gradient is **length** divided by **mass**. It shows the length the spring stretches for every kg added.

e   The line meets the vertical axis at 12 cm. This tells you the length of the spring when there is no mass on it, when it is not stretched.

Practise...   8.2 Real-life graphs

1   Janet is organising a birthday party at a leisure centre for her son. The cost is £50 to book the room plus £5.50 per child for food.

a   Copy and complete this table for the total cost of a party at the leisure centre.

Children	10	20	30	40
Cost (£)	105			

b   Draw a pair of axes with the cost in pounds (£) on the y-axis and the number of children on the x-axis. The y-axis should go from 0 to 300. The x-axis should go from 0 to 40. Plot the points from the table completed in part **a**.

c   Use your graph to find the cost when the number of children is:

i   15

ii   22

iii   31

d   Janet has £200 to spend on a party. What is the maximum number of children her son can invite?

2   Water comes out of a tap at a rate of 125 ml every second.

a   Copy and complete the table for the amount of water coming out of the tap.

Time (seconds)	0	1	2	3	4	5
Water (ml)	0	125				

b   Draw a graph to show this information.

c   Use your graph to find how much water comes out of the tap in 3.2 seconds.

d   How long does it take to get 200 ml of water out of the tap?

e   A kettle holds 1.7 litres of water when full. 1.7 litres = 1700 ml How long will it take to fill the kettle using this tap?

G F E D C

F
E

D

3 This graph shows the amount of petrol used by a car.

a How much petrol is used when the car travels:

  i 60 km

  ii 35 km?

b How many km are travelled when the amount of petrol the car uses is:

  i 4 litres

  ii 2.5 litres?

c Calculate the gradient of the line. What does this tell you about the car?

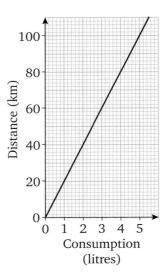

4 Ace Energy supply electricity. They have a standing charge of £8.50 each quarter (this means a customer has to pay £8.50 every 3 months no matter how much electricity they use). They then charge 10.5p per unit of electricity used.

a Copy and complete the table for the cost of electricity from Ace Energy.

Units used	0	500	1000	1500	2000
Cost (£)	8.50	61			

b Draw a graph showing this information.
Betta-supplies have a standing charge of £11.50. They then charge 10p for each unit of electricity used.

c Add a line to your graph showing the cost of electricity from Beta Supplies.

d Which of these electricity suppliers would you advise a new customer to use? Give a reason for your answer.

D
C

5 Budget Pens supply pens to schools. Their prices are shown in the graph. Large orders get a discount.

a What is the total cost of 80 pens from Budget Pens?

b A school buys 80 pens, what is the cost of each pen?

c What is the cost of 250 pens from Budget Pens?

d How many pens does a school have to buy from Budget Pens to receive a discount?

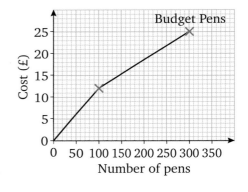

Cheapo Pens are a rival company who also supply pens to schools. They charge 9p for each pen, no matter how big the order.

e Copy and complete the table for the cost of pens from Cheapo Pens.

Number of pens	100	200	300
Cost (£)	9		

f Copy the graph and add a line to show the cost of Cheapo Pens.

g A school wants to order 150 pens. Which company is cheaper and by how much?

h When do Budget Pens and Cheapo Pens charge the same for the same number of pens? How many pens is this and what is the cost?

i A school sells pens to students for 10p each. They use Budget Pens as a supplier. How many pens do they need to sell before they start making a profit?

j A second school buys 300 pens from Budget Pens. How much do they need to sell each pen for so they don't make a loss? Give your answer to the nearest 1p.

6   Tom went for a walk. His walk is represented on this graph.

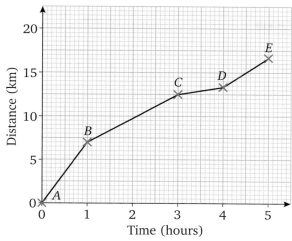

a   In which section of his walk was Tom walking:

  i   fastest

  ii   slowest?

b   What was his average speed in section *BC*?

c   What was his average speed for the whole walk?

7   A small exercise club uses cross trainers, treadmills and exercise bikes. On average a cross trainer burns off 9 calories per minute, and a treadmill 7 calories per minute.

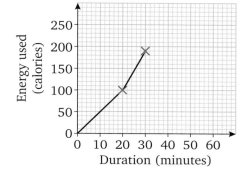

John trains for 30 minutes. The graph shows the calories he burns off. He starts on an exercise bike.

a   How many calories does the exercise bike burn off per minute?

b   What exercise equipment does John use after he has used the exercise bike?

c   Mary uses the cross trainer for 10 minutes then spends 20 minutes on the treadmill. Draw a similar graph for the calories that Mary burns off.

d   Design an exercise programme for Jenny. She wants to burn off over 300 calories, and wants to use all three exercise machines.

Learn... 8.3 Drawing 2-D representations of 3-D objects

Definitions

This diagram shows a cube.

Vertex is the correct name for a corner of the cube. The plural of vertex is **vertices**. A cube has 8 vertices. You can only see 7 in the diagram, there is also one at the back that cannot be seen.

An **edge** joins two vertices. The edges are shown by the lines in the diagram. A cube has 12 edges.

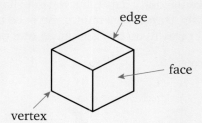

A **face** is the flat part of the cube. A cube has 6 faces. There are three that can be seen in the diagram plus two at the back and one underneath.

Drawing shapes

There are two ways to represent cubes and cuboids in 2-D.

Method 1

This uses special paper, either isometric dotty paper (which just has dots and is sometimes called triangular dotty paper) or isometric paper (with lines).

Using dotty paper

The dots are arranged in triangles 1 cm apart. You need to make sure the paper is the correct way round in order for your diagrams to work.

You should be able to draw triangles like these.

If you can only draw triangles like these

then your paper is the wrong way around.

To draw a cube it is easiest to start with the top. Join the dots to make a rhombus. Then draw in the vertical edges. Finally complete the edges on the base of the cube.

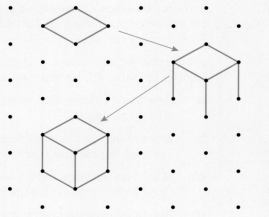

To draw shapes made of cuboids, start with the cube at the front, then work your way back. Remember you cannot see the front face of any cubes that are behind the cube at the front.

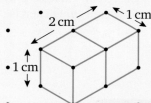

Using isometric paper

This is just like the dotty paper, but the dots have all been joined up. Like dotty paper you need to make sure that it is the correct way round. You need to be able to draw the triangles the same way as for dotty paper. Another way to do this is to make sure there are vertical lines on the page.

These are correct.

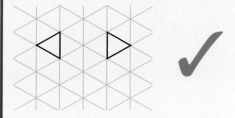

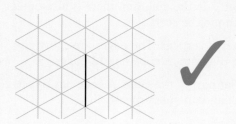

These are not correct.

   ✗   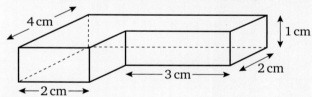   ✗

Method 2

This method involves a sketch without dotty or isometric paper.

Start with a square (this is the front).

Then draw three parallel lines going backwards like this.

Then join the ends of these lines.

Any edges you cannot see need to be shown with dashed lines.

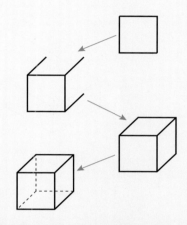

Example: Draw a representation of the following shape on isometric paper.

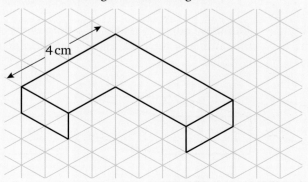

Solution: Make sure your paper is the correct way round. Then start by drawing the top of the shape.

Then start adding the remaining faces.

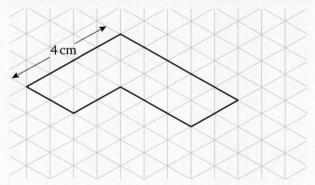

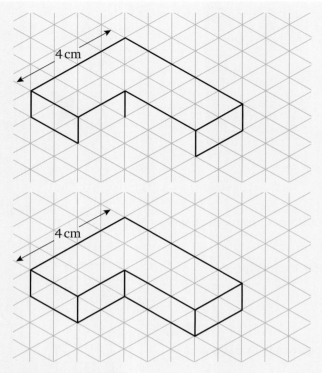

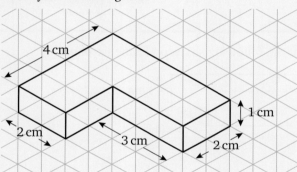

Label your final diagram.

Example: On plain paper, draw a sketch of the cuboid 2 cm by 2 cm by 1 cm, shown here.

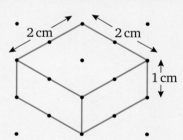

Solution: Start with the front face.

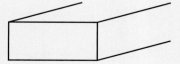

Then add the edges joining the front face to the face at the back.

Join these edges to make the outline of the back that is visible.

Join the vertices to show the edges which are not visible. Remember to use dashed lines for these.

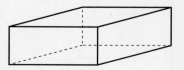

Practise...   **8.3 Drawing 2-D representations of 3-D objects**

G F E D C

E

1 This object is made from four Multilink cubes. If you have 3-D Multilink cubes you can make the object for yourself.

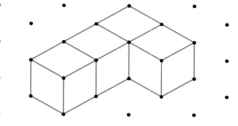

  a Use isometric paper to draw this object. Make sure you have the paper the right way round.

  b Make as many different shapes as you can using four Multilink cubes. Draw each object on isometric paper.

2 On plain paper, draw and label a sketch of a cuboid 3 cm by 3 cm by 2 cm.

3 On plain paper, draw and label a sketch of a cuboid 2 cm by 3 cm by 4 cm.

4 These are drawings of some 3-D shapes. Which drawings are different views of the same shapes? Use Multilink cubes to help you.

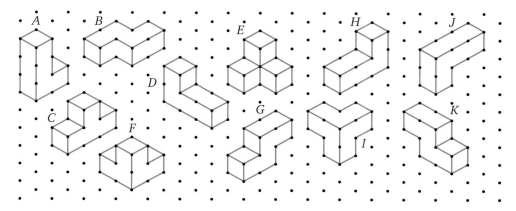

5 Hector has 24 cubes and each is a 1 cm cube. Draw and label a sketch of a cuboid box which will just hold these 24 cubes.

6 **a** Use Multilink cubes to make the shapes shown. Six of the shapes use four cubes and one uses just three cubes. (It may help you to make them using different colours.)

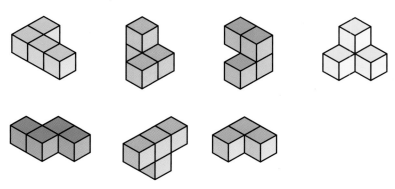

  b Draw each shape on isometric paper.

  c Using your Multilink cubes, fit the shapes together to make a cube. This is called the Soma cube and was invented by a Danish mathematician, Piet Hein.

7   **a**   Use Multilink cubes to make the following shapes.
(It may help you to make them using different colours.)

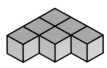

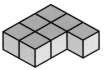

b   Draw each shape on isometric dotty paper.

c   Fit the shapes together to make a cube. This cube is called the Diabolical cube and was 'invented' by Professor Louis Hoffman. It was included in his book *Puzzles Old and New* in 1893.

8   **a**   Make these two shapes from Multilink cubes. (It may help you to make them using different colours.)

b   Each of these drawings is an outline of a solid made from these two shapes.

Copy the outlines onto isometric dotty paper. Draw in the missing lines. Shade in the two shapes to show how they fit together to make the solid.

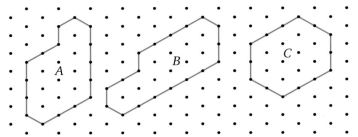

9   **a**   Using five Multilink cubes make as many different shapes as you can.

b   Draw each of them on isometric dotty paper.

c   How can you tell you have found them all?

d   Repeat parts **a** and **b** with six Multilink cubes.

> **Hint**
> You will find it easier to answer this if you work systematically.

Learn...   **8.4 Plans and elevations**

3-D objects may be viewed from different directions.

The view from above is called the **plan** view.

The view from the front is called the front **elevation**.

The view from the side is called the side elevation.

Sometimes the front elevation will look the same as the side elevation.

Lines which cannot be seen are drawn using dashed lines.

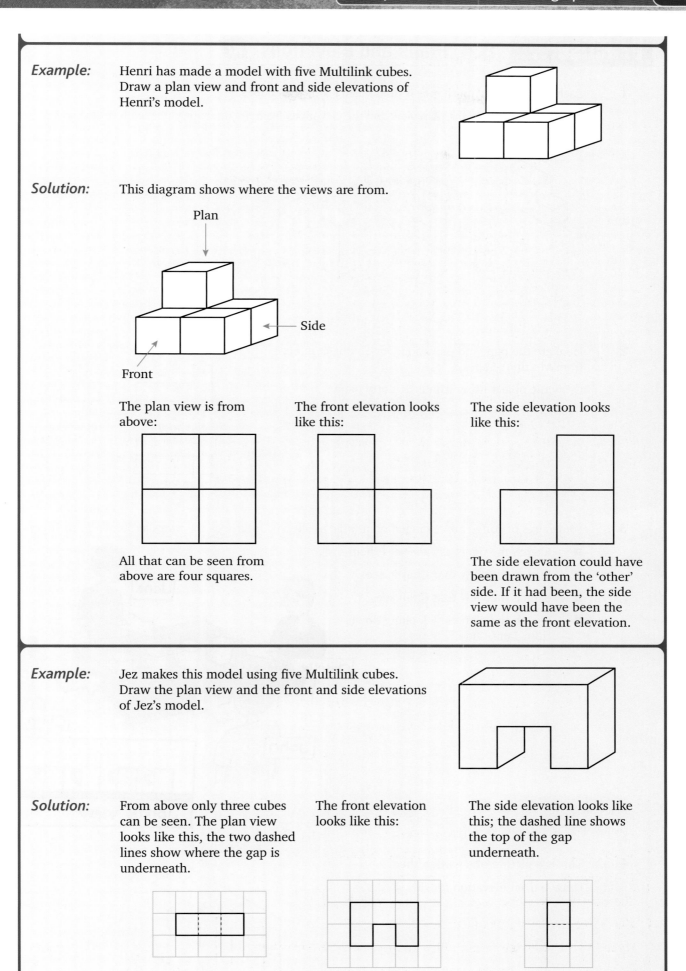

Example: Henri has made a model with five Multilink cubes. Draw a plan view and front and side elevations of Henri's model.

Solution: This diagram shows where the views are from.

Plan

Side

Front

The plan view is from above:

All that can be seen from above are four squares.

The front elevation looks like this:

The side elevation looks like this:

The side elevation could have been drawn from the 'other' side. If it had been, the side view would have been the same as the front elevation.

Example: Jez makes this model using five Multilink cubes. Draw the plan view and the front and side elevations of Jez's model.

Solution: From above only three cubes can be seen. The plan view looks like this, the two dashed lines show where the gap is underneath.

The front elevation looks like this:

The side elevation looks like this; the dashed line shows the top of the gap underneath.

Practise... 8.4 Plans and elevations

1 Each of the following shapes is made using Multilink cubes.

For each shape draw the plan view and the elevations from the directions labelled F (front) and S (side).

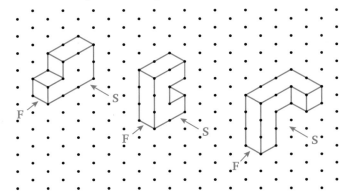

2 Here are the plan and elevations for an object made from Multilink cubes.

Draw the object using isometric dotty paper.

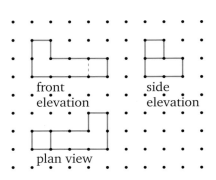

front elevation   side elevation

plan view

3 Four mugs of coffee are arranged on a table as shown. John sees the mugs as shown in the diagram.

a Sketch the mugs as Jane sees them.

c Sketch the view that Charlie sees.

b Sketch the view that Chim sees.

d Sketch the plan view looking down from Jane's side.

John's view

4 This is a picture of a garden shed.

Draw a front elevation for this shed.

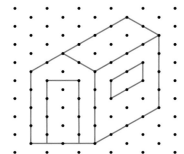

D

5   The following sketches show some pieces of furniture from a doll's house.

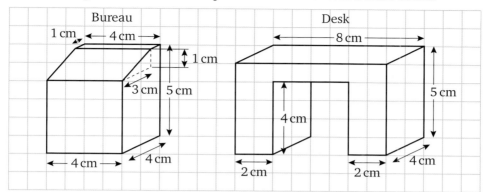

a   Ian drew these plan views of the furniture. What mistakes has Ian made?

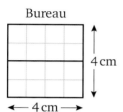

b   Draw the plan views accurately and label the dimensions of the desk and bureau.

6   Alan, Bridgette, Charlotte and Dan are doing some work on plans and elevations in a mathematics lesson.

They have arranged two books on the table. They are sitting around the table as shown in the diagram.

a   They each draw the front elevation that they see. Which of the following does each draw?

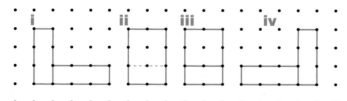

b   They rearrange the books as shown.

Draw the front elevation that each person sees. Label them clearly.

 **Learn...** **8.5 Nets** **k!**

A **net** is a flat shape that can be made into a 3-D shape when you fold it up.

Example: Draw the net of an open cube.

Solution:

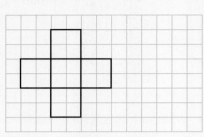   If you fold the sides up you get an open box.

    to make

Example: Draw the net of a cuboid.

Solution: This is the net of a cuboid. It does not have an open top.

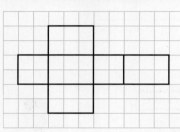

> **Hint**
>
> If you wish to cut out a net and glue it together to make a solid object you will need to add 'tabs'. Be careful to add the correct number of tabs.

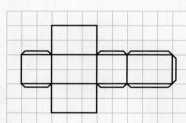

Practise... **8.5 Nets** **k!**   **G F E D C**

G

1   Which of the following are nets of a cube?

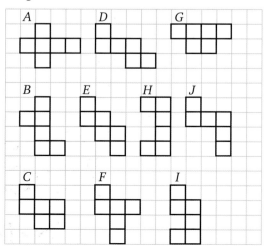

F

2   The following diagrams show two views of the delivery box for a computer.

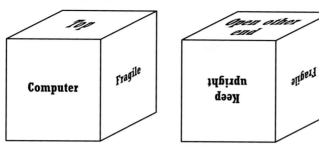

This is the net of the box.
Copy and complete the labelling of the box.

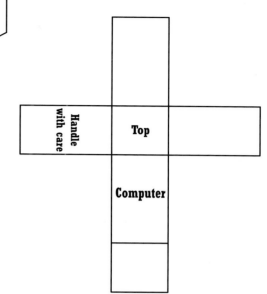

E

3   This is the sketch of a solid.

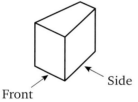

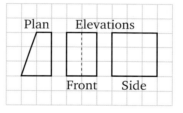

a   Use squared paper to draw an accurate net of the solid.

b   Add tabs to your net, cut it out and glue it together to make a model of the solid.

4   A tetrahedron has four faces, each of which is an equilateral triangle.

a   Use triangular dotty paper, or isometric paper, to draw the net of a tetrahedron. Each of the four equilateral triangles should have sides 4 cm long.

b   Add tabs and cut out your net and stick it together to make a tetrahedron.

5   The diagram shows a triangular prism.

Three of the faces are squares and two are equilateral triangles.
Which of the following could be nets of the triangular prism?

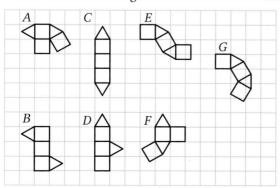

E

6 The diagram shows a square-based pyramid.

It has a square base, and four identical isosceles triangles as faces.

Max drew this net for the square-based pyramid.

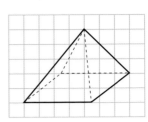

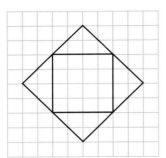

a What is wrong with this net?

b Draw a net that does make a square-based pyramid.

E
D

7 Mike uses card to make a gift box for his mother's birthday present as shown.

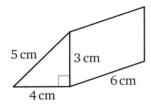

5 cm   3 cm   6 cm   4 cm

a Draw an accurate net of the gift box on squared paper.

b What is the area of the card Mike uses?

c What is the volume of the gift box when it is made?

⚠ 8 This is a regular octahedron. It has eight identical faces, each of which is an equilateral triangle.

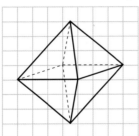

Draw the net of an octahedron on dotty paper.
Add tabs and make the octahedron to check it works.

⚠ 9 Which of the following nets could be a net of this cube?

A   *C*   *E*

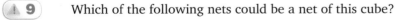

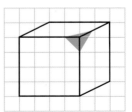

B   *D*   *F*

    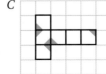   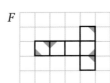

10 James has bought his mother a present for Mother's Day. It is in a cylindrical pack as shown.

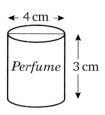

James wants to make a gift box in the shape of a pyramid that is large enough to contain the cylinder.

Design a gift box for James. Draw the net and make the gift box.

11 **a** Draw the net of a cube with an open top. It is an example of a polyomino.

b Because it is made from five squares it is called a pentomino. Find all the different pentominoes. Which ones are the net of an open box?

c A hexomino is made of six squares. Find all the possible hexominoes. How can you be sure that you have found them all? Which ones are the net of a cube?

8 **Assess** (k!)

1 This is a map of Pirate Island.

Write down the coordinates of:

a Sharp Point

b Pirate Falls

c High Mountain

d both ends of the Ancient Wall

e Smugglers' Cave.

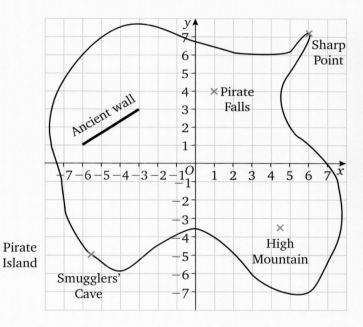

Pirate Island

2 The diagram shows a cuboid.

Draw the net of this cuboid accurately on squared paper.

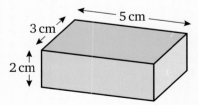

3 This is the net of a cube.

A dice has the numbers 1 to 6 on its faces. The numbers on faces opposite each other add up to 7. So 1 is opposite 6.

Copy this net and put the numbers 1 to 6 on it.

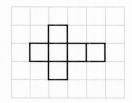

F

4 This is the net of a 3-D shape. It is made from three identical rectangles and two equilateral triangles.

What is the name of the shape it makes?

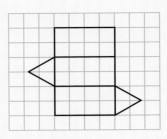

E

5 Draw a sketch of a cuboid which is 4 cm long, 3 cm high and 3 cm wide.

F
E

6 A car salesman is paid a monthly salary plus a bonus for each car he sells. His pay is shown in the graph.

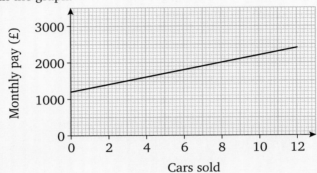

a What is his basic monthly salary?

b How much does he receive if he sells:
  i 6 cars   **ii** 5 cars?

c How much does he get paid for each car he sells?

d How many cars does he need to sell if his income is to be £2800 in one month?

e Calculate how much he will earn if he sells 22 cars in one month.

E

7 **a** Write down the equations of the lines A, B and C.

b Draw a pair of x- and y-axes from −5 to 5 on squared paper.
Draw and label lines with the following equations:
  i $x = -2$
  ii $y = -2$

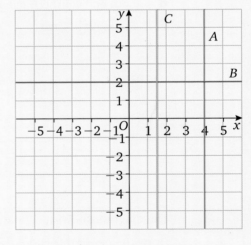

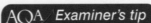 **AQA** *Examiner's tip*

Always label the x- and y-axes. When you draw a line on the grid remember to label it with its equation.

8 This diagram shows three points, A, B and C.
Each point is the vertex of a parallelogram.
Write down the coordinates of the three possible positions of the fourth vertex.

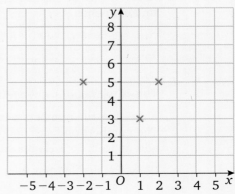

D

9 Linda arranges two books as shown.

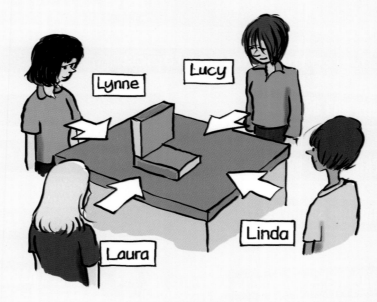

a Draw four diagrams showing the views that Linda, Laura, Lynne and Lucy see.

b Draw a plan view of the arrangement.

AQA Examination-style questions

1 The graph shows Adil's bicycle journey.

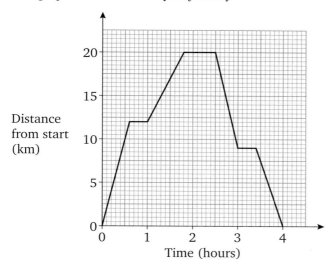

a How many times does Adil stop on his journey? *(1 mark)*

b How many times is Adil exactly 10 km from the start of his journey? *(1 mark)*

c What is the total distance that Adil travels on his journey? *(1 mark)*

d Calculate Adil's average speed during the first 30 minutes of his journey. Give your answer in kilometres per hour. *(2 marks)*

AQA 2007

You have covered the following topics:

Fractions and decimals

Angles

Working with symbols

Percentages and ratios

Perimeter and area

Equations

Properties of polygons

Coordinates and graphs

All these topics will be tested in this chapter and you will find a mixture of problem solving and functional questions. You won't always be told which bit of maths to use or what type a question is, so you will have to decide on the best method, just like in your exam.

Example: This is a regular hexagon.

It has a perimeter of 18 centimetres.

Not drawn accurately

Three of these hexagons are used to make this shape.

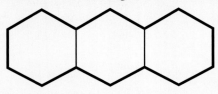

Not drawn accurately

Hint

The two sides inside the shape are not part of the perimeter.

Work out the perimeter of the shape.

(3 marks)

Solution: As a diagram is given, you can count the number of sides.

The number of sides = 6

Each side of the regular hexagon = 18 cm ÷ 6

= 3 cm

The shape has 14 sides.

Perimeter of shape = 14 × 3 cm

= 42 cm

$\overline{AQA}$ *Examiner's tip*

Mark each side of the diagram as you count so that you get the total correct.

Mark scheme

- 1 mark for attempting 18 ÷ 3 or obtaining the side length as 3 cm.
- 1 mark for multiplying 14 by the side length.
- 1 mark for the correct final answer of 42 cm.

$\overline{AQA}$ *Examiner's tip*

You are allowed a calculator in Unit 3 so use it to work out 14 × 3

Example: The diagram shows a right-angled triangle.

a   Work out the value of *x*.

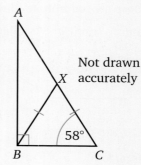

Not drawn accurately

(2 marks)

b   *X* is a point on *AC* so that *BX = CX*

The triangle is cut along the line *BX* to make two triangles.

Show that triangle *ABX* is an isosceles triangle.

Not drawn accurately

(2 marks)

Solution: **a**   Start by thinking about what you know about the angles of a triangle.
The angles of a triangle add up to 180°.
A right angle is 90°.

So the angle at *A* must be   *180° − 90° − 58° = 32°*

x = 32°

b   *BX = CX*. Think about what this tells you about the triangle *BXC*.
This means **triangle *BXC* is isosceles.**
Think of what you know about isosceles triangles.
Two sides are the same length and the base angles are the same.

So **angle *XBC* must also be 58°.**

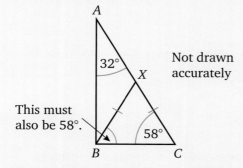

This must also be 58°.

Not drawn accurately

> **AQA   *Examiner's tip***
>
> When a question says 'Show that…', you must make sure that you set your working out very clearly so that you do not miss out any steps in your working.

> **AQA   *Examiner's tip***
>
> It can help to mark the angles on the diagram.

Because angle *B* is 90°, angle *ABX* must be **90° − 58° = 32°**

Triangle *ABX* has two angles which are the same, at *A* and *B*.

So this triangle must also be isosceles.

Mark scheme

- 1 mark for the method of working out angle *A*.
- 1 mark for the final answer 32°.
- 1 mark for seeing that angle *XBC* is 58°.
- 1 mark for working out angle *ABX* as 32° and explaining why this means the triangle is isosceles.

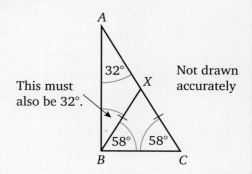

This must also be 32°.

Not drawn accurately

Consolidation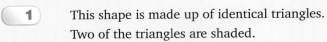

G

1   This shape is made up of identical triangles.
Two of the triangles are shaded.

   a   How many more triangles must be shaded
so that $\frac{3}{4}$ of the shape is shaded?

   b   Which of these percentages is the same as $\frac{3}{4}$?

      25%   34%   43%   75%

2   Choose a word from the list that describes each of these angles.

   acute   obtuse   reflex   right

   a   **b**   **c**   **d**

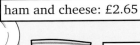

3   Margaret goes to a shop to buy sandwiches for her friends at work.

   This is her list of what to buy.

   2 ham and cheese sandwiches

   5 chicken sandwiches

   4 bags of crisps

   She has a £20 note in her purse.
Will she have enough money?
You **must** show all your working.

chicken: £2.85

ham and cheese: £2.65

crisps: 40p

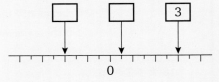

4   Name each shape. Pick from the list given each time.

   a

      rectangle   hexagon   trapezium   pentagon   square

   b

      oblong   parallelogram   trapezium   square   rectangle

G
F

5   **a**   Copy the number line. Fill in the two empty boxes
with the numbers represented on the number line.

   b a, b, and c are three different numbers.

      $a \times b$ is negative.

      $b \times c$ is positive.

      Which of the following statements, A, B and C, is correct?

      A: $a \times c$ is positive.

      B: $a \times c$ is negative.

      C: $a \times c$ may be positive or negative.

      You **must** give a reason for your answer.

6   Joss and Hayley are finding points on this grid.

Joss draws a cross on points where $x + y = 6$

Hayley draws a circle around points where $y - x = 2$

Here is the grid after they have each had two goes.

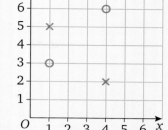

a   If they continue finding points, how many points on this grid will have a circle drawn around them?

b   Write down the coordinates of the point that Joss could plot that is on the x-axis.

c   Write down the coordinates of the point that Hayley could plot that is on the y-axis.

d   How many points will have a cross with a circle drawn around it?
Write down the coordinates of these points.

7   Shapes A and B are drawn on a square grid.

The area of shape A is 20 cm^2.

Work out the area of shape B.

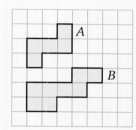

8   In the diagram:

a is the smallest angle

b is 10° bigger than a

c is 20° bigger than b

d is 30° bigger than c.

Work out the size of angle a.

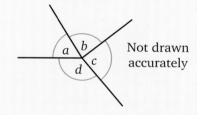

Not drawn accurately

9   The perimeter of a regular pentagon is 64 cm.
Work out the length of each side.

10   **a**   Davinda buys 40 carpet tiles.
One-fifth of the tiles are blue and the rest are white.
Blue tiles cost £4.15 each.
White tiles cost £2.75 each.
Work out the total cost of the tiles.

b   Each tile is a square of side 30 cm.
Davinda arranges the 40 tiles in a 10 by 4 rectangle on the floor.
Work out the perimeter of the rectangle that he has made.

11   Jack is 1.2 kg heavier than Kylie.

Together they weigh 18 kg.

Work out Jack's weight.

12   A wall in a room is 330 centimetres wide.

Su Ling wants to hang two pictures on the wall.

She wants the distances between the pictures and the ends of the wall to be the same.

Each picture is 45 centimetres wide.

Work out the distance, d, between the pictures.

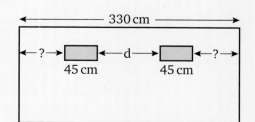

13

a Simplify fully $\frac{14d}{7}$

b $s = -5$ and $u = 4$
Work out the value of $2s - 3u$

c Simplify $7g + 4f - g - 6f$

14 The diagram shows the plan for a new park.

The local council says that the percentage of the area allocated to each purpose is:

gardens 52%

children's play area 4%

sports pitches 34%

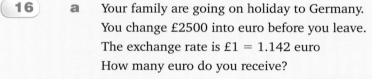

gardens

sports pitches

children's play area

Not drawn accurately

a Erin says 'There is a mistake with these figures'.

Give a reason why Erin knows this.

b The area of the park is 0.675 km².

i The local council says the figure for the area of the gardens should have been 62%.
What is the area for the gardens?
Give your answer to three decimal places.

ii In a change to the plan, 0.054 km² of the park is turned into a car park.
What percentage of the park is this?

15 Two of the angles of an isosceles triangle are x and $2x$.
Work out two possible values for x.

16

a Your family are going on holiday to Germany.
You change £2500 into euro before you leave.
The exchange rate is £1 = 1.142 euro
How many euro do you receive?

b During your holiday you change another £360 into euro.
The exchange rate is now 1 euro = £0.90
How many euro do you receive?

c At the end of your holiday you want to change 150 euros into pounds.
The exchange rate in Germany is £1 = 1.165 euro
The exchange rate in Britain is £1 = 1.172 euro
In which country will you get more pounds for your 150 euro?
Give a reason for your answer.

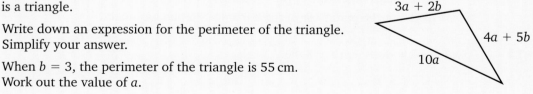

S1865085

0859757

17 Here is a triangle.

a Write down an expression for the perimeter of the triangle.
Simplify your answer.

b When $b = 3$, the perimeter of the triangle is 55 cm.
Work out the value of a.

$3a + 2b$

$4a + 5b$

$10a$

18 Work out the angles marked with a letter.

a

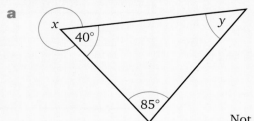

x

$40°$

y

$85°$

b

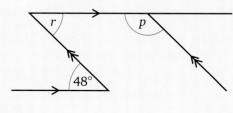

r

p

$48°$

Not drawn accurately

19 **a** Round these numbers to one significant figure.

      i   8.256

      ii 19.2

      iii 7821

   b Use your answers to part **a** to estimate $\dfrac{19.2 \times 7821}{8.256}$

20 Here are four sketch graphs.

Graph A   Graph B   Graph C   Graph D

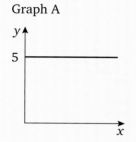 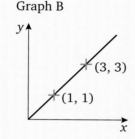 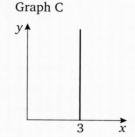 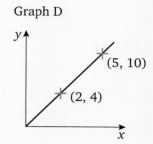

Pick the equation from those given below that each graph represents.

   $y = 3$ $y = 2$ $y = x$ $x = 3$

   $y = 5$ $x = 5$ $y = \frac{1}{2}x$ $y = 2x$

21 The diagram shows the cross-section of a roof.

The two sides of the roof are perpendicular.

One side of the roof slopes at 35° to the horizontal.

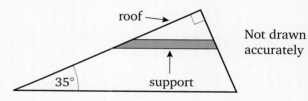

A horizontal support is put in.

The support is made from a rectangular piece of wood.

To make it fit, the wood is cut along the dotted lines.

Not drawn accurately

Work out the values of x and y.

22 The boat *Clarabel* breaks down and sends out a distress call.

The call is heard by two other boats, *Aramis* and *Bellamy*.

They set out to intercept *Clarabel*.

Aramis sets out on a bearing of 055°.

Bellamy sets out on a bearing of 260°.

Copy the diagram and mark the position of *Clarabel*.

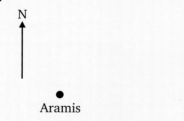

23 Which of these shapes has the largest area?

Circle   Triangle

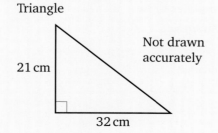

Not drawn accurately

You **must** show your working.

D

24 Ahmed has the following test results.

French: 61%

ICT: 75 out of 120

In which subject did he get the better score?

You **must** show your working.

25 Alice is four years younger than Ben.

She is nine years older than Carly.

The total age of all three people is 79.

How old is Alice?

26 A rectangle has a length that is twice the width.

Four of these rectangles make this shape.

Work out the perimeter of the shape.

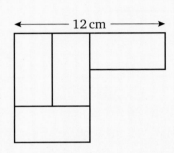

D
C

27 An activity centre organises climbing trips for the public.

a Each member of staff at the centre must not take more than 8 members of the public in their group.

A party of 76 people book a trip.

12 members of staff are available to take the party.

i Are there enough members of staff to take all the people?
You **must** show all your working.

ii What is the ratio of staff to people for this trip?
Give your answer in the form $1 : n$

b In 2009 there were 18 726 people who went on climbing trips.

In 2010 this number increased to 19 871.

Work out the percentage increase.

Give your answer to one decimal place.

Bump up your grade

To get a Grade C you should be able to work out a percentage increase or decrease.

C

28 **a** Simplify the following expression.

$6(x - 2) - 2(2x + 3)$

b Two numbers in the following expression have been covered up.

$\bullet x - \bullet(4 - x)$

The expression simplifies to $5x - 8$.

Work out the two numbers.

29 Here is part of a number line.

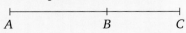

$AB : BC = 4 : 3$

If $A = 2.8$ and $C = 14.7$, work out the value of B.

30 The points *P*, *Q* and *R* are on a straight line.

PQ : *QR* is 3 : 1

Work out the coordinates of point *Q*.

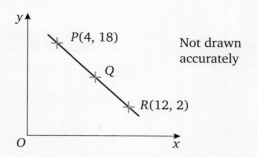

Not drawn accurately

31 An interior angle of a regular polygon *A* is 135°.

Another regular polygon *B* has two sides fewer than polygon *A*.

Work out one of the interior angles of polygon *B*.

AQA Examination-style questions

1 **a** *PQR* is a straight line.

Find the value of *x*.

Not drawn accurately

(2 marks)

b The three lines shown below meet at a point.

Find the value of *y*.

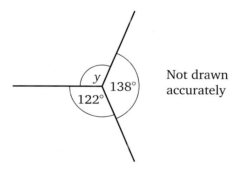

Not drawn accurately

(2 marks)

c In the diagram, *AB* is parallel to *CD*.

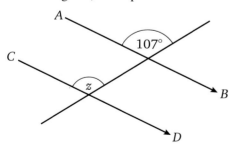

Not drawn accurately

Find the value of *z*.

(1 mark)

AQA 2006

2 Complete this table.

Expression	Value
2*x*	8
5*x*	
2*x* + 3*y*	5
y	
3*x* − *y*	

(2 marks)

AQA 2006

9 Reflections, rotations and translations

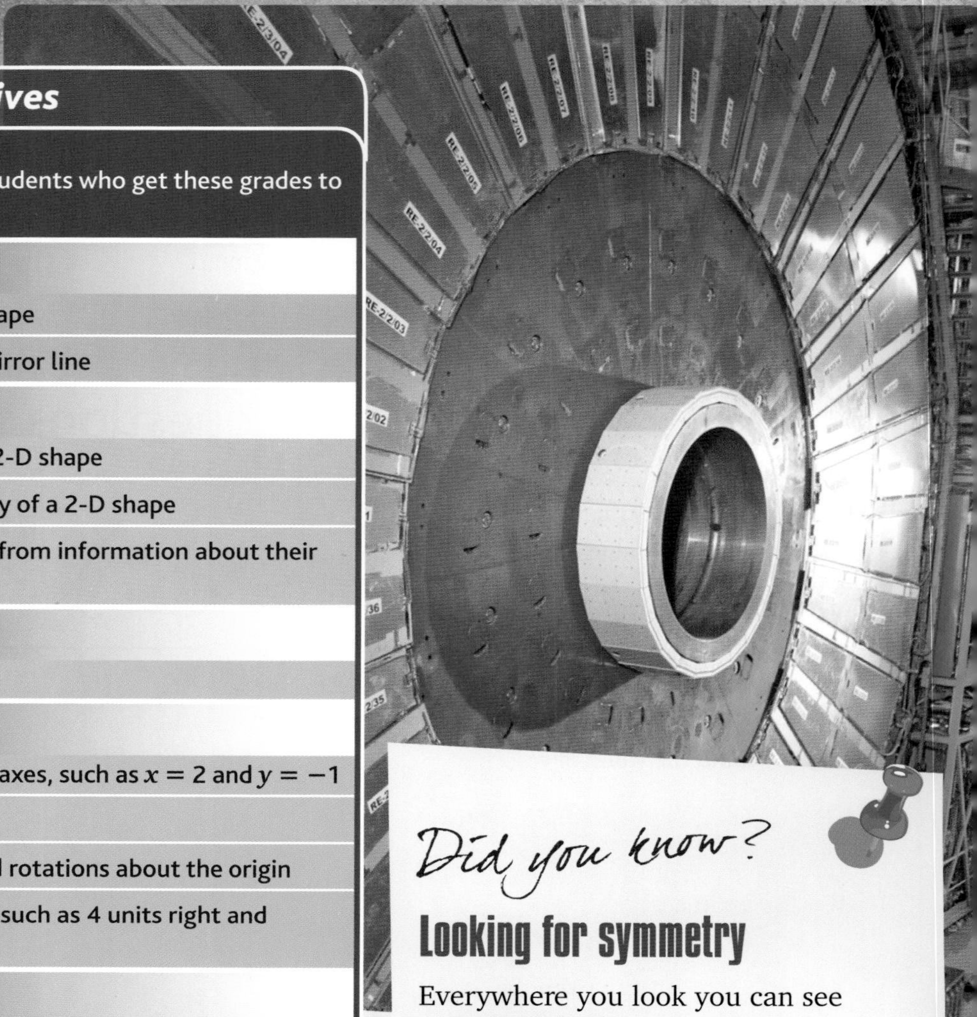

Objectives

Examiners would normally expect students who get these grades to be able to:

G

draw a line of symmetry on a 2-D shape

draw the reflection of a shape in a mirror line

F

draw all the lines of symmetry on a 2-D shape

give the order of rotational symmetry of a 2-D shape

name, draw or complete 2-D shapes from information about their symmetry

E

reflect shapes in the axes of a graph

D

reflect shapes in lines parallel to the axes, such as $x = 2$ and $y = -1$

rotate shapes about the origin

describe fully reflections in a line and rotations about the origin

translate a shape using a description such as 4 units right and 3 units down

C

reflect shapes in lines such as $y = x$ and $y = -x$

rotate shapes about any point

describe fully reflections in any line parallel to the axes, $y = x$ or $y = -x$ and rotations about any point

find the centre of a rotation and describe it fully

translate a shape by a vector such as $\begin{pmatrix} 4 \\ -3 \end{pmatrix}$.

Did you know?

Looking for symmetry

Everywhere you look you can see symmetry, both in nature and in man-made constructions. This is the Compact Muon Solenoid, one of the detectors in the Large Hadron Collider. This is the world's largest and highest-energy particle accelerator. It is a symmetrical construction being used to advance science. After being built by scientists and engineers from over 100 countries, it was in operation for the first time in September 2008.

Key terms

reflectional symmetry	rotational symmetry
reflection	order of rotation
line of symmetry	centre of rotation
mirror line	clockwise
object	anticlockwise
image	angle of rotation
congruent	vector
vertex, vertices	mapped
coordinates	translation
rotation	

You should already know:

- ✔ how to plot positive and negative coordinates
- ✔ equations of lines, such as $x = 3$, $y = -2$, $y = x$ and $y = -x$
- ✔ names of 2-D and 3-D shapes
- ✔ that angles in a full turn equal 360°.

 **Learn...   9.1 Reflection** *k!*

Reflectional symmetry

One half of this shape is a mirror **reflection** of the other. It has one **line of symmetry**. The **mirror line** is drawn so one side of the shape is a reflection of the other.

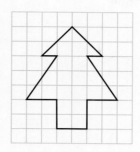 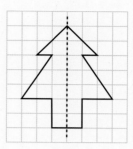

This shape has 3 lines of symmetry. One half of the shape can be reflected in three different ways.

Example: How many lines of symmetry does each grid pattern have?

Draw in the lines of symmetry that you find.

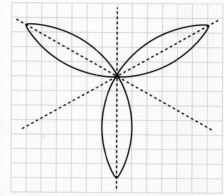

Solution: Grid pattern **a** has two lines of symmetry.
Grid pattern **b** has no lines of symmetry.

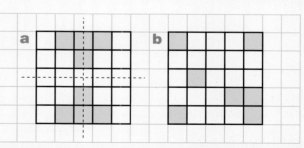

Reflecting a shape in a given line

Shapes can be reflected on grids, as shown in the previous examples, and on the axes on a graph.

The shape *T* has been reflected in the *x*-axis.

The equation of the *x*-axis is $y = 0$

The value of *y* is 0 for every point on the line.

The shape *T* is called the **object** and the reflection *T'* is called the **image**.

The object and the image are **congruent**. Two shapes are congruent if they are exactly the same size and shape.

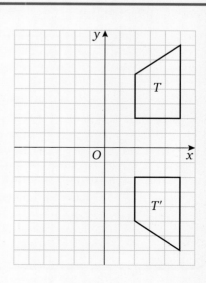

This shape, P, has been reflected in the line $x = 2$

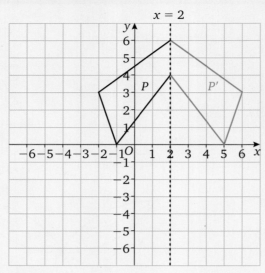

The reflection of P is P'.

The **vertices** or corners of P' can be written as **coordinates**.

P' has coordinates $(2, 4)$, $(2, 6)$, $(5, 0)$ and $(6, 3)$.

Example: Step a: Draw a pair of x- and y-axes from -8 to 8.

Step b: Draw a polygon, K, by plotting and joining the points $(-2, 1)$, $(2, 1)$, $(0, 4)$

Step c: What is the name of this polygon?

Step d: Reflect the polygon in the line $y = -2$. Label your reflected shape K'.

Step e: Write down the coordinates of the vertices of the image, K'.

Solution:

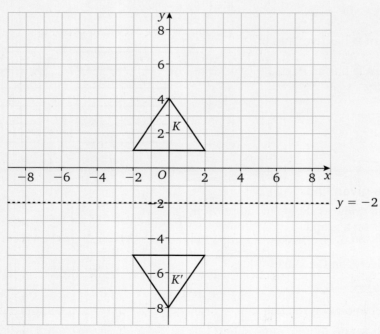

The polygon is an isosceles triangle.

K' has coordinates $(-2, -5)$, $(2, -5)$, $(0, -8)$.

Practise... 9.1 Reflection

G

1   **a**   How many lines of symmetry does each letter have?
You can use tracing paper to help you.

i   **M**   iii   **H**   v   **X**

ii   **A**   iv   **B**   vi   **T**

b   Write down six letters that have no lines of symmetry.

2   Copy each shape onto squared paper. Draw its image after being reflected in the mirror line.

a 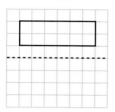   **c** 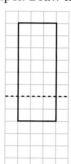   **d** 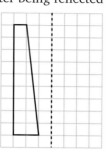   **f**

b 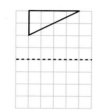   **e**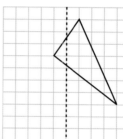

3   **a**   All these shapes have reflectional symmetry.
How many lines of symmetry has each shape got?
Copy the shapes and draw all the lines of symmetry on your diagrams.
You can use tracing paper to help you.

i   iii   v   vii   ix

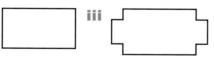

ii   iv   vi   viii

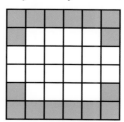

b   Do all triangles have the same number of lines of symmetry? Give a reason for your answer.

F

4   **a**   Give the number of lines of reflectional symmetry for each grid pattern.

i 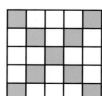   iii    v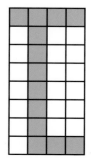

F

ii 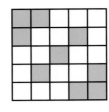 iv 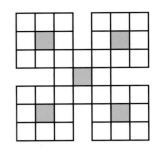 vi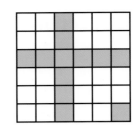

b On squared paper, draw your own grid diagram with 4 lines of symmetry.

5 Copy each diagram onto squared paper and complete it to give the required number of lines of symmetry.

a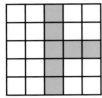

4 lines of symmetry

b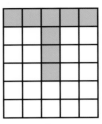

2 lines of symmetry

c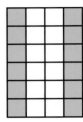

0 lines of symmetry

E

6 **a** **i** Copy each diagram onto axes on squared paper.

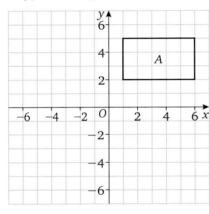

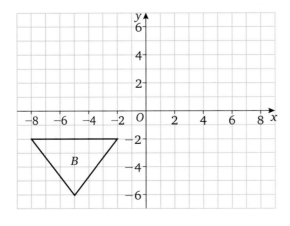

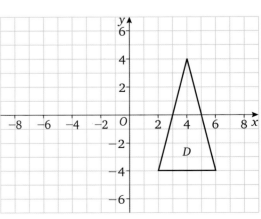

ii Reflect each shape in the *x*-axis. This will give you an image of the shape.

iii Write down the coordinates of the vertices of each shape.

iv Write down the coordinates of the vertices of each image.

v   What is the relationship between the coordinates of the vertices of the shape and its image?

> **Hint**
>
> Vertex means corner. Vertices are the plural of vertex.

b   **i**   Copy each diagram in question part **a i** onto axes on squared paper.

  ii   Reflect each shape in the *y*-axis.

  iii   Write down the coordinates of the vertices of each shape.

  iv   Write down the coordinates of the vertices of each image.

  v   What is the relationship between the coordinates of the vertices of the shape and its image?

7   **a**   **i**   Draw a pair of *x*- and *y*-axes from −6 to 6.

  ii   Draw a polygon, *P*, by plotting and joining these points:

  (3, 3), (5, 1), (3, −4), (1, 1)

  iii   What is the mathematical name of the polygon?

  iv   Reflect the polygon *P* in the *y*-axis and write down the coordinates of the vertices of the image, *P′*.

  v   Reflect the polygon *P* in the *x*-axis and write down the coordinates of the vertices of the image, *P″*.

> **AQA** *Examiner's tip*
>
> You need to give the special name of the polygon. For example, don't just answer 'triangle'. You need to say if it is an equilateral triangle, isosceles triangle, right-angled triangle or scalene triangle.

b   **i**   Draw a pair of *x*- and *y*-axes from −6 to 6.

  ii   Draw a polygon, *Q*, by plotting and joining these points:

  (1, 2), (5, 2), (3, −1), (−1, −1)

  iii   What is the name of the polygon?

  iv   Reflect the polygon *Q* in the *y*-axis and write down the coordinates of the vertices of the image, *Q′*.

  v   Reflect the polygon *Q* in the *x*-axis and write down the coordinates of the vertices of the image, *Q″*.

8   For each diagram **a**–**d**, write down the equation of the line of reflectional symmetry.

a

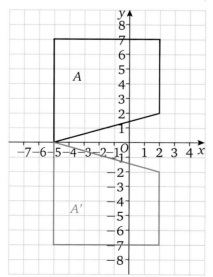

c

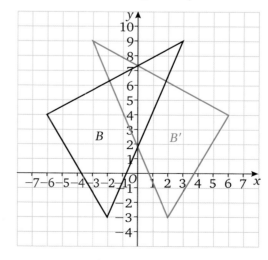

b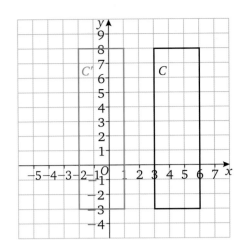

d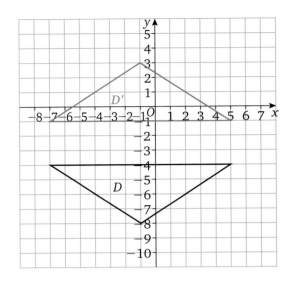

9

a Draw a pair of x- and y-axes from −6 to 6.

b Draw a polygon, R, by plotting and joining these points:
(1, −1), (3, −1), (3, −3), (1, −3)

c What is the name of the polygon?

d Reflect the polygon in the line x = y and write down the coordinates of the vertices of the image, R'.

e Reflect the polygon in the line x = −y and write down the coordinates of the vertices of the image, R".

f Write down what you notice about the coordinates.

> **Bump up your grade**
>
> To get a Grade C you need to know how to reflect in the lines y = x and y = −x as well as horizontal and vertical lines.

10

a This diagram shows a shape, A, and its reflection, A'. Describe the reflection that maps A onto A".

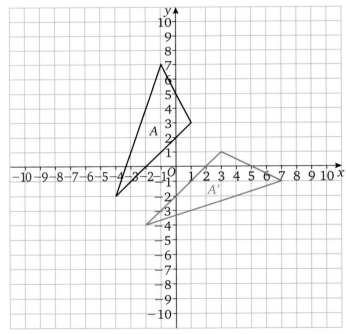

b Write down the coordinates of the vertices of both triangles.

c Find a rule that connects the coordinates before and after the reflection.

d Use your rule to work out the coordinates of the vertices of the image of the polygon which has the coordinates given below. Use the same line of reflection. Try and work it out by following your rule and not by drawing:
(2, 3), (−1, −2), (0, −4), (2, −2)

e Can you find a rule for each of these lines of symmetry without drawing them?

   i x = 0   **ii** y = 3   **iii** x = −5

11   Alisha is investigating the reflectional symmetry of quadrilaterals. She has to find out which quadrilaterals have reflectional symmetry and whether all quadrilaterals have the same number of lines of symmetry.

    a   Make a list of all the special quadrilaterals.

    b   Draw each quadrilateral with the lines of symmetry.

    c   Write an answer for Alisha's investigation.

12   This pattern is being created using repeated reflection.

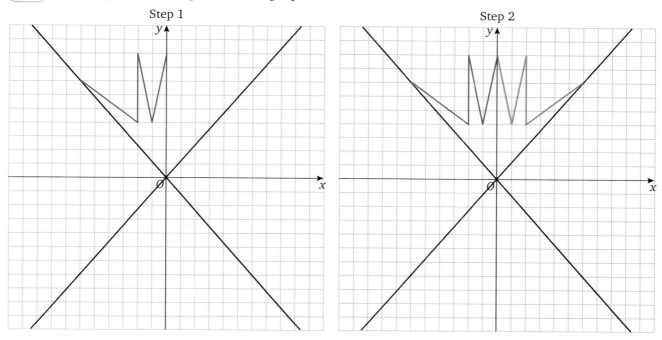

Step 1   Step 2

    a   Write down the equations of all the lines of reflection.

    b   Copy and complete the pattern.

    c   Design your own pattern using the same lines of reflection.

Learn... 9.2 Rotation

Rotation means turning.

Rotational symmetry

How many ways does a shape look the same while it is being turned through 360°?

When turned around the point of rotation this shape looks the same in four positions.

It has **rotational symmetry of order 4 about the centre**.

When turned around the point of rotation this shape looks the same in one position.

It has **rotational symmetry of order 1 about the centre**.

You can always check your answer using tracing paper.

Trace the shape onto tracing paper.

Mark the centre of the shape.

Mark a small arrow or number one facing the top of the page.

This is the first position.

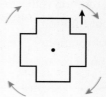

Hold your pencil on the centre point and turn your tracing clockwise.

As you turn the shape, count the number of times your traced image looks the same as the starting image before you get back to the starting point.

This is the **order of rotational symmetry** of the shape.

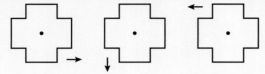

Example: Find the order of rotational symmetry of this shape.

Solution: When it is rotated around the centre this octagon looks the same in eight places.
The octagon has rotational symmetry of order 8.

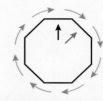

Rotating a shape about a given point

Shapes can be rotated on grids or axes.

The **centre of rotation** is the fixed point around which the object is rotated. It is given using coordinates.

The amount the shape turns is given as an angle or fraction of a complete turn, e.g. 270° or $\frac{3}{4}$ turn.

The direction of rotation is given as **clockwise** or **anticlockwise**.

This shape has a centre of rotation $(-1, 2)$.

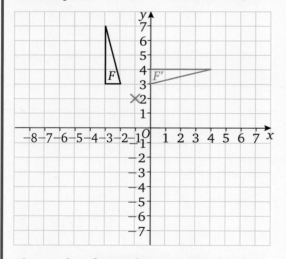

The **angle of rotation** is 90° clockwise.

The shape is called the **object** and the rotation is called the **image**.

The object and the image are **congruent**. If the object is F then the image is F′.

To describe a rotation fully, you must give:

- the centre of rotation
- the angle of rotation
- the direction of rotation.

AQA Examiner's tip

In your exam, you can ask for tracing paper to find the centre of rotation.

Example: In this diagram, the right-angled triangle, A, has been rotated anticlockwise by 90° about the origin.

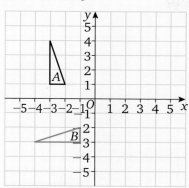

The image of A following this rotation is B.

a   Write down the clockwise rotation that also maps A onto B.

b   Draw a diagram showing the image of A after a rotation of 90° clockwise about the point (−1, 0). Label the image C.

c   What are the three coordinates of the vertices of C?

Solution: **a**   A clockwise rotation of 270° about the origin gives the same image as an anticlockwise rotation of 90° about the origin.

b

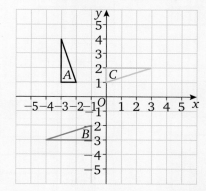

c   The coordinates of the vertices of C are: (0, 1), (0, 2), (3, 2)

Practise... 9.2 Rotation

G F E D C

1   **a**   Write down the order of rotational symmetry of each of these letters. You can use tracing paper to help you.

i   **H**   iii   **N**   v   **X**

ii   **A**   iv   **S**   vi   **Z**

b   Harry says the order of rotational symmetry of the letter O is 8. Is Harry correct? Give a reason for your answer.

F

F

2　**a**　Write down the order of rotational symmetry of each of these shapes.

You can use tracing paper to help you.

i 　iii 　v 　vii

ii 　iv 　vi 　viii

b　Is the order of rotational symmetry of a triangle always the same?
Draw a sketch of the special triangles and use these to explain your answer.

c　Is the order of rotational symmetry of a pentagon always the same?
Sketch some pentagons and use these to explain your answer.

AQA *Examiner's tip*

Make sure you use the word 'rotation' and not the word 'turn'.

3　What is the order of rotational symmetry for each of these grid shapes?

a 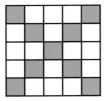　**c** 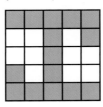　**e**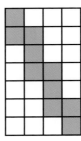

b 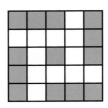　**d** 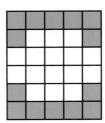　**f**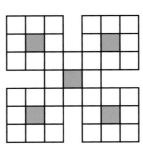

4　Copy and complete the grid diagrams so each one has the given order of rotational symmetry.

a 　　**b** 　　**c** 　　**d**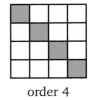

order 2　　　　order 2　　　　order 1　　　　order 4

5　The hands of this clock each turn through 360° in one rotation.
The hour hand turns through an angle of 180° from 12.00 to 6.00.

Write down the angle the **hour** hand turns clockwise through from 12.00 for each of these times.

a　3.00　　**c**　9.00　　**e**　8.00

b　11.00　　**d**　4.00

6   **a**   For each question part, write down the order of rotational symmetry and the angle of rotation.

i 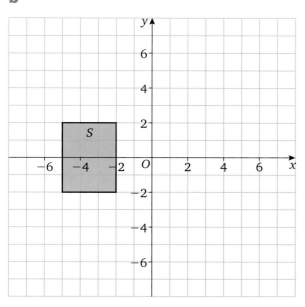   **ii**   **iii**

b   If a shape turns through 60° and looks the same as in its original position. What is the order of rotational symmetry?

7   Draw x- and y-axes from −7 to 7 for each question part.

Copy each object R to W and rotate it by 180°.

Use the origin as the centre of rotation.

Label each image R′ to W′.

a

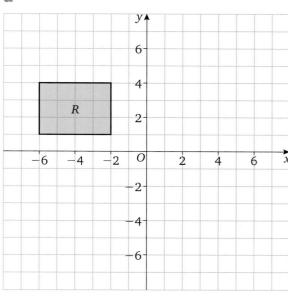

c

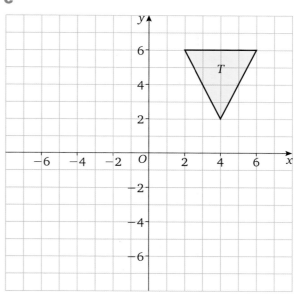

b

d

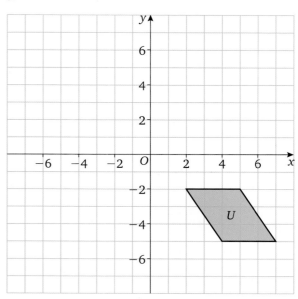

D

e

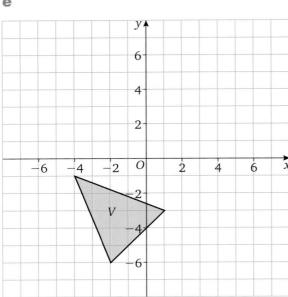

f

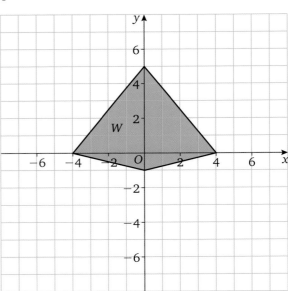

8 **a** Draw a pair of x- and y-axes from −8 to 10 for each question part.

Copy each object A to F and rotate it by 90° clockwise.

Use the point (1, 2) as the centre of rotation.

Label each image A' to F'.

b Now rotate each image A' to F' by 270° clockwise.

Use the point (−1, −1) as the centre of rotation.

Label each image A' to F".

i

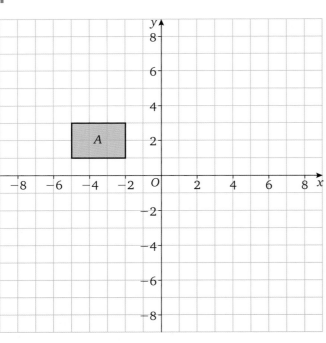

ii

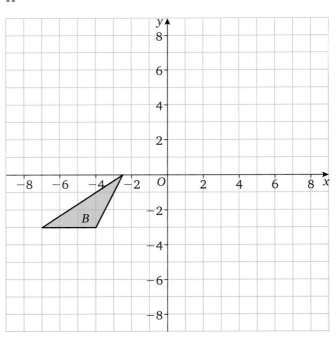

iii

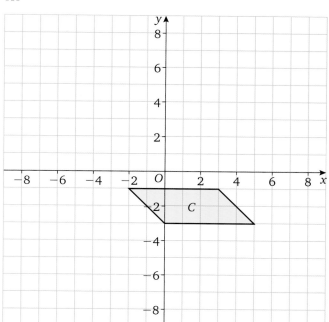

v

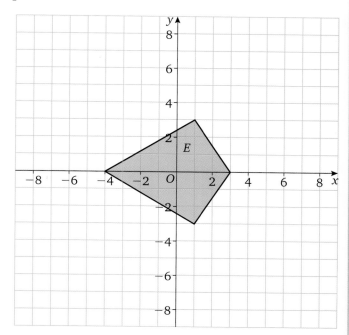

iv

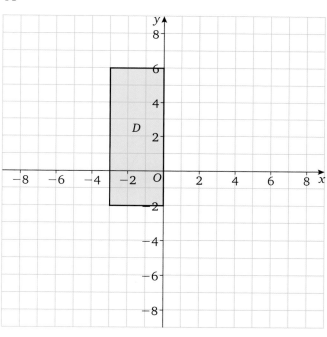

vi

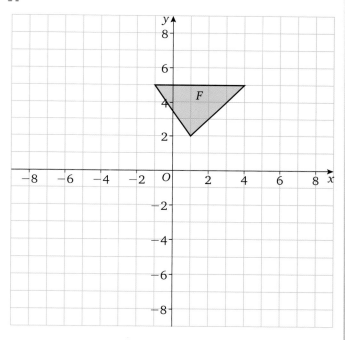

D

9 **a** Draw a pair of *x*- and *y*-axes from −8 to 8 for each question.

Copy each object *G* to *L* and rotate it by 90° anticlockwise.

Use the point (−1, −1) as the centre of rotation.

Label each image *G′* to *L′*.

b Now rotate each image *G′* to *L′* by 270° anticlockwise.

Use the point (0, −2) as the centre of rotation.

Label each image *G″* to *L″*.

Bump up your grade

To get a Grade C you need to be able to rotate objects around any point and not just the origin.

D

i

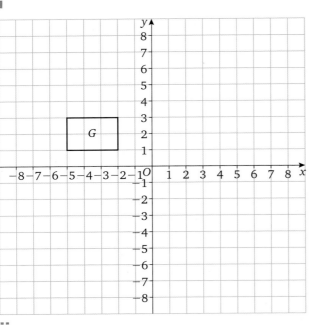

ii

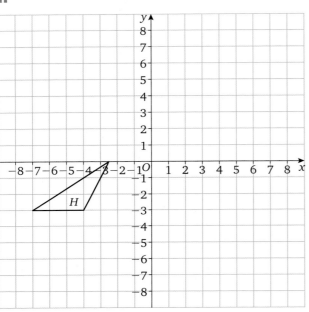

iii

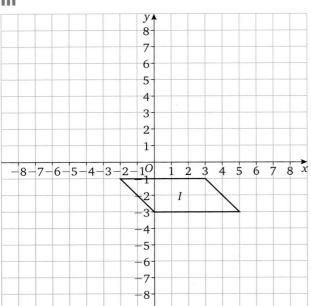

iv

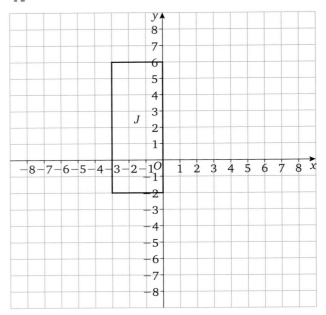

v

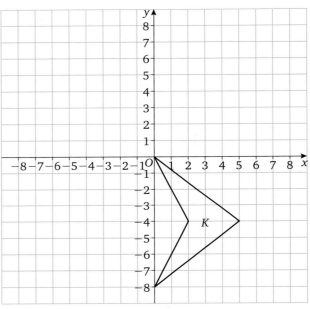

vi

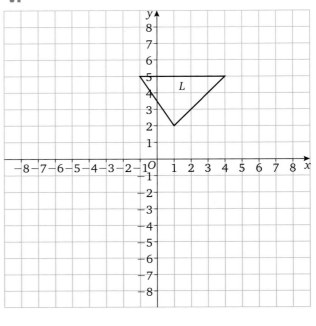

10

a Draw a pair of x- and y-axes from −7 to 7.

b Draw a triangle T by plotting and joining these points: (0, 0), (−2, −1), (−1, −2)

c Draw the image V by rotating T by 90° clockwise about (0, 0).

d Draw the image W by rotating T by 180° about (0, 0).

e Draw the image X by rotating T by 270° clockwise about (0, 0).

11

a Draw a pair of x- and y-axes from −7 to 7.

b Draw a polygon Q by plotting and joining these points: (4, 1), (6, 1), (4, −1), (3, −1)

c Draw the image R′ by rotating Q by 90° clockwise about (2, 3).

12 For each diagram **a–f**, give the coordinates of the centre of rotation as A is mapped onto B.

a

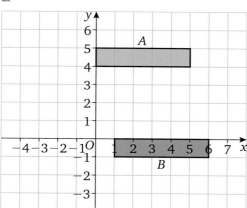

d

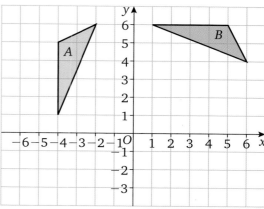

b

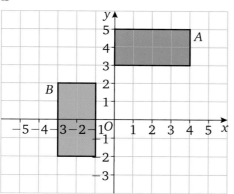

e

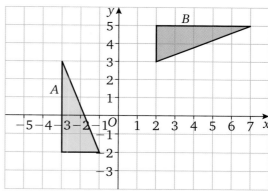

c

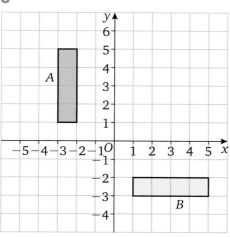

f

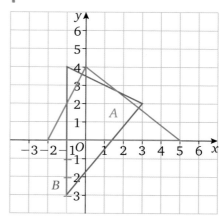

Hint

Use tracing paper to try different centres of rotation until you find the one that works for both images.

13 This diagram shows two different images of a triangle *A* after two rotations by the same angle.

B is the image of *A* after a rotation clockwise about the point (0, 2).

C is the image of *A* after a rotation anticlockwise about point (0, 2).

Find the position of triangle *A*.

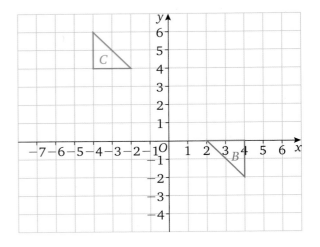

14 Rotational symmetry is often used in the design of company logos.

a The arrows on this logo make it look like it has rotational symmetry of order 2.

Explain why this logo does not have rotational symmetry of order 2.

b This logo uses arrows too.
  i What is the order of rotational symmetry of this logo?
  ii Find the angle of rotation.

c **i** Design your own company logo with rotational symmetry.
  ii What is the order of rotational symmetry of your logo?

15 Use squared paper to draw your own grid shapes with the following orders of rotational symmetry.

a 2   **b** 4   **c** 1

16 **a** **i** Draw a line 4 cm long. Mark one end of the line as the centre of rotation.
  ii Using a protractor, measure an angle of 45° from this centre of rotation and draw another 4 cm line.
  iii Continue rotating the line and drawing a new line, always at 45° from the previous line.

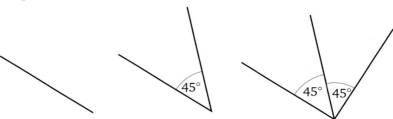

b What is the order of rotational symmetry of your final pattern?

c If you repeat this using a 30° angle instead of 45°, what is the order of rotational symmetry?

d Why is it impossible to draw a rotational symmetry pattern with 38° as the angle of turn?

e Is it possible to draw a rotational symmetry pattern with a 72° angle of turn? Give a reason for your answer.

Learn... **9.3 Translation**

This shape has been translated.

Every point moves the same distance in the same direction.

The **object** and the **image** are **congruent**.

The distance and direction can be written as a **vector**.

Shape A has been **mapped** onto shape B by a **translation** of 2 to the right and 3 units up.

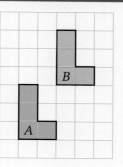

Written as a vector, this is $\binom{2}{3}$ ← 2 to the right ← 3 up

The vector that maps shape C onto shape D is $\binom{0}{-4}$

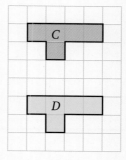

The top number is the horizontal move.
- If the number is positive the shape moves to the right.
- If the number is negative the shape moves to the left.

The bottom number is the vertical move.
- If the number is positive the shape moves up.
- If the number is negative the shape moves down.

Vectors can be used on any grid with or without a pair of axes.

Example: Describe the translation that maps shape A onto shape B in each diagram.

a Diagram 1

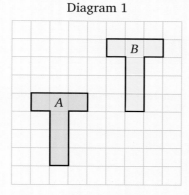

b Diagram 2

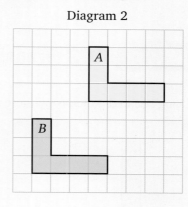

Use vectors in your answers.

Solution: **a** A has moved 4 to the right and 3 up so the vector translation is $\binom{4}{3}$

b A has moved 3 to the left and 4 down so the vector translation is $\binom{-3}{-4}$

Example: **a** Copy the diagram, drawing triangle T with the coordinates (2, 4), (4, 4) and (4, 1).

b Draw the image of T after the vector translation $\binom{-5}{-3}$. Label the image R.

c Write down the coordinates of the vertices of R.

d What is the relationship between the coordinates of the vertices of T and the coordinates of the vertices of R?

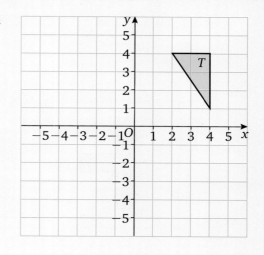

Solution: **a/b**

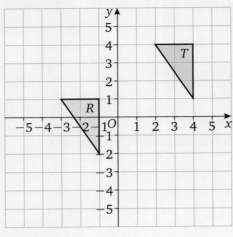

c   (−3, 1), (−1, 1), (−1, −2)

d   The x-coordinates of R are all 5 less than the x-coordinates of T.
The y-coordinates of R are all 3 less than the y-coordinates of T.

Practise...   9.3 Translation 🔊   G F E D C

D

1   Using squared paper, copy these shapes and translate each one by the given amount.

a

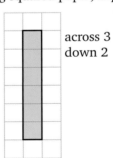

across 3
down 2

d

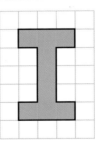

across 2
up 5

b

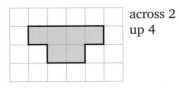

across 2
up 4

e

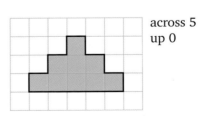

across 5
up 0

c

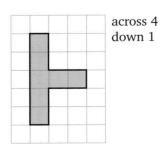

across 4
down 1

f

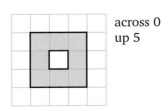

across 0
up 5

C

2 Describe the translation that maps shape *A* onto shape *B* in each diagram.
Give your answers as vectors.

a
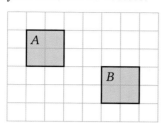

d
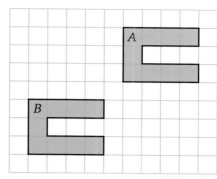

b
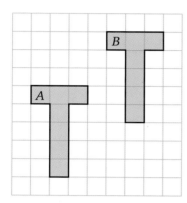

e
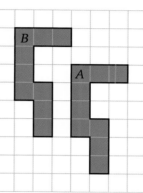

c
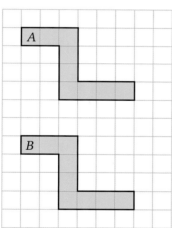

f
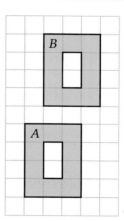

Bump up your grade

To get a Grade C use vectors to describe translations rather than right/left and up/down.

3 **a** Look at the diagram and then write down the vector translation that maps:

  i   *A* onto *B*

  ii *A* onto *C*

  iii *B* onto *C*

  iv *C* onto *B*.

b Compare your answers for question parts **iii** and **iv**. Write down what you notice.

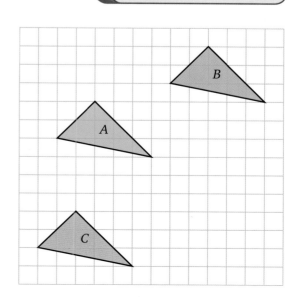

C

4   Look at the diagram and then write down the vector translation that maps:

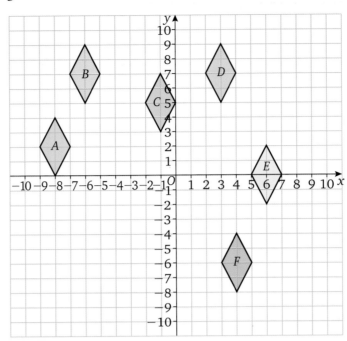

a   A onto B   c   E onto D

b   C onto F   d   B onto E.

5   a   i   Draw x- and y-axes from −5 to 5.

ii   Plot and join the points (−4, −5), (−1, −5), (−4, 2). Label this shape T.

b   Translate T using the vector translation $\begin{pmatrix} 4 \\ 3 \end{pmatrix}$ to give the image U.

c   Translate U using the vector translation $\begin{pmatrix} -5 \\ 0 \end{pmatrix}$ to give the image V.

d   Translate V using the vector translation $\begin{pmatrix} 6 \\ -1 \end{pmatrix}$ to give the image W.

e   Describe fully the single translation that maps T directly onto W.

6   a   Draw a pair of x- and y-axes from −5 to 5 onto squared paper.

b   Begin at the origin as a starting point. Translate this point using the vectors below. Each translation follows the previous one. After each translation, put a cross at the new point.

$$\begin{pmatrix} 2 \\ 1 \end{pmatrix} \quad \begin{pmatrix} 1 \\ 1 \end{pmatrix} \quad \begin{pmatrix} 1 \\ 0 \end{pmatrix} \quad \begin{pmatrix} 1 \\ -1 \end{pmatrix} \quad \begin{pmatrix} 0 \\ -2 \end{pmatrix} \quad \begin{pmatrix} -2 \\ -2 \end{pmatrix} \quad \begin{pmatrix} -3 \\ -2 \end{pmatrix}$$

c   Join the crosses you have drawn in the order in which you drew them.

d   Reflect the shape in the y-axis.

7   Some wallpaper designs use patterns that have been translated.
The design is made by repeatedly translating the feature design horizontally and vertically.

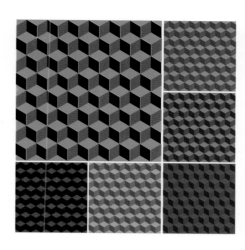

Emma is designing a wallpaper pattern on squared paper.

She translates her design by the vector $\begin{pmatrix} 4 \\ 3 \end{pmatrix}$

She then repeats the pattern in a different colour.

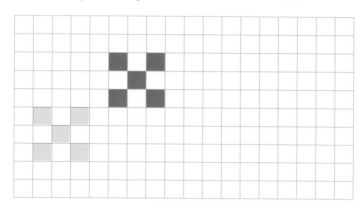

a   On squared or isometric paper, create a simple design.

b   Choose a translation vector. Repeat your pattern using your chosen translation vector to create your own wallpaper design.

8   Sam and Holly designed this board for a vector game.

A **vector route** starts at X and lands on each of the other squares before returning to X.

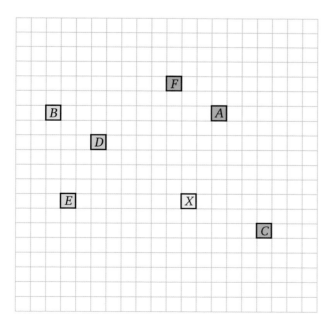

a   Holly wrote down this vector route.
Write down the order in which she landed on the squares.

$$\begin{pmatrix} 5 \\ -2 \end{pmatrix} \quad \begin{pmatrix} -13 \\ 2 \end{pmatrix} \quad \begin{pmatrix} -1 \\ 6 \end{pmatrix} \quad \begin{pmatrix} 3 \\ -2 \end{pmatrix} \quad \begin{pmatrix} 5 \\ 4 \end{pmatrix} \quad \begin{pmatrix} -3 \\ -2 \end{pmatrix} \quad \begin{pmatrix} -2 \\ -6 \end{pmatrix}$$

b   Sam chose to visit the squares in this order: $X\,F\,D\,C\,B\,A\,E\,X$
Write down Sam's vector route.

c   Create a vector route of your own and test it on a friend.

9 Assess *k!*

G

1 How many lines of symmetry does each letter have?

     a **K**   b **G**   c **E**   d **T**   e **R**

2 Copy the diagrams and draw the image of each shape reflected in the mirror line.

   a    b

F

3   a Copy this grid pattern and draw on the lines of symmetry.

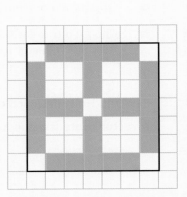

   b Copy this grid pattern and shade seven more squares so the grid has rotational symmetry of order 4 about its centre.

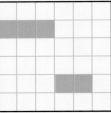

   c Copy this grid pattern and shade four more squares so the grid has 2 lines of reflectional symmetry.

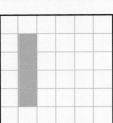

4   a What is the order of rotational symmetry of this shape?

   b What is the angle of rotation?

5

a **i** Draw a pair of x- and y-axes from -6 to 6.

  ii Draw the shape A by plotting and joining these points:
  $(-1, 1), (-3, 1), (-4, 4), (-1, 3)$

b What is the name of polygon A?

c Reflect A in the line $y = -1$. Label this image B.

d What are the coordinates of the vertices of image B?

E

6

a Copy the diagram on x- and y-axes labelled -5 to 5.

b Rotate the shape 90° clockwise about the origin.

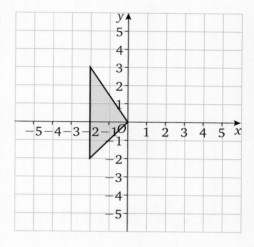

D

7 Describe the translation that maps shape A onto shape B.

Use vectors to give your answer.

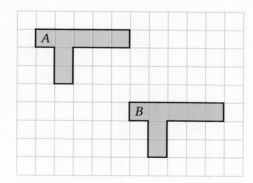

C

8 What is the equation of the line of reflectional symmetry in this diagram?

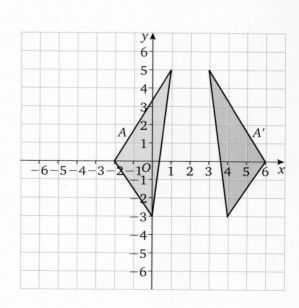

C

9

a   Copy the diagram.

b   Write down the coordinates of the vertices of the triangle A.

c   Reflect the triangle in the line $x = -y$. Label the image B.

d   Write down the coordinates of the vertices of triangle B.

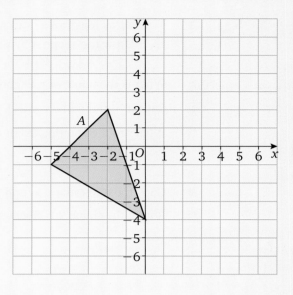

AQA Examination-style questions

1   Triangles A, B and C are shown on the grid.

    a   Describe fully the single transformation that maps triangle A onto triangle B.   *(3 marks)*

    b   Write down the vector which describes the translation of triangle A onto triangle C.   *(1 mark)*

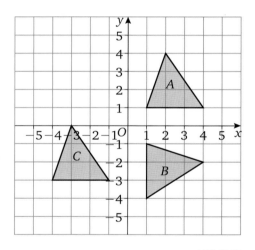

AQA 2009

10 Formulae

Objectives

Examiners would normally expect students who get these grades to be able to:

G

use a formula in words such as:
Total pay = rate per hour × number of hours + bonus

F

substitute positive numbers into a simple formula such as $P = 2L + 2W$

use formulae from mathematics and other subjects such as $v = u + at$

E

use formulae such as $P = 2L + 2W$ to find W given P and L

substitute negative numbers into a simple formula such as $F = 1.8C + 32$

derive formulae such as $C = 35h + 55$

D

substitute numbers into more complicated formulae such as $C = \dfrac{(A + 1)D}{9}$

derive more complex formulae

distinguish between an expression, an equation and a formula.

Key terms

formula
symbol
subject
expression

equation
substitute
value

Did you know?

Falling down

The following formula shows how far a body will fall under gravity if air resistance is ignored:

$$h = \tfrac{1}{2}gt^2$$

h is the vertical distance travelled
g is the acceleration of gravity on the Earth's surface
t is the length of time the body falls.

Aristotle believed that if two objects were dropped, the heavier one would fall faster (and travel further in a given time). By dropping cannonballs from the leaning tower at Pisa, Galileo showed that all objects will travel the same distance at the same rate regardless of the mass of the body.

You should already know:

✔ the order of operations (BIDMAS)

✔ the four rules applied to negative numbers

✔ how to simplify expressions by collecting like terms

✔ how to solve linear equations

✔ how to find fractions of a quantity.

10.1 Writing formulae and expressions using letters and symbols

A **formula** tells you how to work something out such as Area of a triangle = half base × height.

A formula may be written in words or **symbols**.

For example, the above formula for area could be written as $A = \frac{1}{2} \times b \times h$ or $A = \frac{b \times h}{2}$

When you write formulae using symbols you need to remember:

- If a stands for a number then $2 \times a$ can be written as $2a$.
- Always write the number in front of the letter.
- The expression $3x + 5$ means multiply x by 3 and then add 5.
- The expression $5(x - 2)$ means subtract 2 from x and then multiply the answer by 5.

Sometimes you are asked to write your formula in its simplest terms. This means you need to collect all the like terms together.

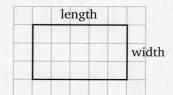

AQA **Examiner's tip**

Be careful when you choose your own letters in problems, as some letters are easily confused with numbers.

Z and 2 can get confused

I or l and 1 can get confused

b and 6 can get confused

q and 9 can get confused

S and 5 can get confused

Example: Find a formula for the perimeter of this rectangle.

length

width

Solution: Using the formula, perimeter = distance around the rectangle
= length + width + length + width

So perimeter = 2 × length + 2 × width

Or, perimeter = 2 × (length + width)

This is an example of a formula in words.

Example: Find a formula for the perimeter, P, of this rectangle.

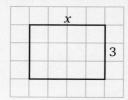

x

3

Solution: In this rectangle, length = x and width = 3

So perimeter = $x + 3 + x + 3 = 2x + 6$

Or, perimeter = 2 × length + 2 × width   Or, perimeter = 2 × (length + width)

$P = 2x + 2 \times 3$ $P = 2(x + 3)$

$P = 2x + 6$ $P = 2x + 6$

This is an example of a formula in symbols.

Note

In the last example, we have a formula for P, which enables us to find the perimeter for any value of x.

P is the **subject** of the formula.

The right-hand side of the formula is an algebraic **expression**.

The formula can also be used to find x when the perimeter is known.

If the perimeter is 21, there is just one value of x and it is found by solving the **equation**: $21 = 2x + 6$

AQA **Examiner's tip**

Make sure you understand the difference between an equation (which can be solved) a formula (into which values are substituted) and an expression (which does not have an equals sign).

Example: This shape is made from two rectangles. All the lengths are in cm.

Find a formula for its perimeter. Give your formula in its simplest form.

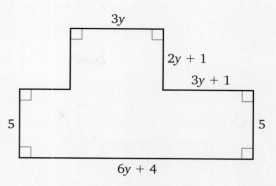

Solution: The first thing you need to do is to find the lengths of any unknown sides.

From the diagram the bottom side of length $6y + 4$ must be equal to the three top sides added together.

Adding the two known lengths (coloured blue) gives $3y + 3y + 1 = 6y + 1$

The length of the side coloured green must therefore be 3, as $3 + 6y + 1 = 6y + 4$

The side coloured red must be $2y + 1$, as it 'matches' the side opposite.

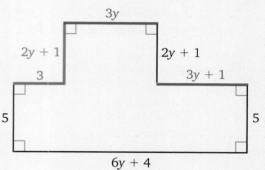

(Remember, the question says the shape is made from two rectangles.)

The perimeter of the shape is the distance around the outside.

Perimeter, $P = 3y + 2y + 1 + 3y + 1 + 5 + 6y + 4 + 5 + 3 + 2y + 1$

To give the formula in its simplest form you need to add the y terms together, and add the number terms together.

$P = 3y + 2y + 3y + 6y + 2y + 1 + 1 + 5 + 4 + 5 + 3 + 1$

$P = 16y + 20$

> ### AQA *Examiner's tip*
> Make sure you use all of the sides when finding the formula for the perimeter of a shape. You should go around the shape carefully starting from one corner. A common error is to miss out a side with no length marked.

10.1 Writing formulae and expressions using letters and symbols

 Practise...

G F E D C

1   Write down a formula using words for the area of this rectangle.

G

length
width

2   Write down a formula for the area, A, of this rectangle.

F

x
3

F

3 Find a formula for the total, T, of the angles in this triangle.

Write your formula in its simplest form.

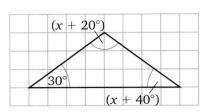

4 Write a formula using symbols for this word formula.

Multiply π by the diameter to find the circumference of a circle.

F E

5 The width of a rectangle is w cm.
The length of the rectangle is 6 cm longer than the width.

Write down an expression for the length of the rectangle.

6 Four students are asked to write down a formula for the perimeter of a rectangle with sides y cm and $2y - 3$ cm.

Andy writes down $3y - 3 \times 2$
Cara writes down $P = 2y + 2y - 3$
Lisa writes down $6y - 6$
Pete writes down $P = 2(y + 2y - 3)$

Which student is correct?

Can you explain the mistakes made by the other students?

E

7 James is making a square enclosure for his pet gerbil.
It is made from a piece of cardboard x cm long.

Write down a formula for the length, s, of each side of the square enclosure.

8 Rachel is paid £5 for every hour she helps with jobs around the house.
She also receives £10 pocket money every month.
In any month her total income from pocket money and jobs is £T.

Work out T in a month when she works for:

a   0 hours   **c** $3\frac{1}{2}$ hours   **e** h hours.

b   2 hours   **d** $5\frac{1}{4}$ hours

9 Pete the plumber is paid a £45 call-out fee plus £25 for every hour he spends working on a repair.

How much is Pete paid for a job that lasts:

a   1 hour   **c** $4\frac{1}{2}$ hours   **e** h hours?

b   3 hours   **d** $2\frac{3}{4}$ hours

D

10 Find a formula for the perimeter of this shape.

Give your formula in its simplest form.

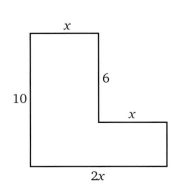

⚠11 Jane is a builder. She gets paid £20 for each hour she works plus a bonus of £25 for each hour of overtime she works. Choose appropriate letters and write down a formula for her total pay in any week.

Learn... 10.2 Substitution

When you **substitute** into a formula you replace the letters or words in the formula with **values**.

You need to be careful when you calculate using the numbers you are substituting into the formula. You can use your calculator to help you avoid making errors.

When you substitute numbers into a formula you get an **equation**. This equation can then be solved to find the value of the unknown letter. Remember you have to do the same to both sides when you solve equations.

> **AQA Examiner's tip**
>
> When you substitute into a formula always write the formula down first, and then rewrite it replacing the letters with the values given. There will be a mark for doing this in your exam.

Example: This formula is used to work out the pay, P, for a plumber in pounds.

$P = 55 + 30h$ where h is the number of hours worked.

a   What is the call-out charge?

b   How much is the plumber paid for working:

  i 2 hours   **ii** 3 hours 15 minutes   **iii** 20 minutes?

c   How many hours does the plumber work in order to be paid £75?

Solution: **a**   The formula is $P = 55 + 30h$

When $h = 0$, $P = 55$. The call-out charge is for turning up but not doing any work.

The call-out charge is £55.

> **AQA Examiner's tip**
>
> Remember to include the units in your answer.

b   **i**   For 2 hours work $h = 2$

$P = 55 + 30h$ Copy the formula.

$P = 55 + 30 \times 2$ Substitute (replace the letter with its value).

$P = 55 + 60$ Calculate carefully (use BIDMAS).

$P = 115$

The plumber is paid £115.

ii   3 hours 15 minutes = 3.25 hours (remember there are 60 minutes in 1 hour.

15 minutes = $\frac{15}{60}$ hour = $\frac{1}{4}$ hour = 0.25 hour)

$P = 55 + 30h$ Copy the formula.

$P = 55 + 30 \times 3.25$ Substitute (replace the letter with its value).

$P = 55 + 97.5$ Calculate carefully (use BIDMAS).

$P = 152.5$

The plumber is paid £152.50.

> **AQA Examiner's tip**
>
> When you work with money on your calculator it will often miss off the final zero. You must write money correctly in your exam. You will not gain the final mark for answers such as £152.5 so remember to write £152.50.

iii $h = 20$ minutes (remember there are 60 minutes in 1 hour.

20 minutes = $\frac{20}{60}$ hour = $\frac{1}{3}$ hour)

$P = 55 + 30h$ Copy the formula.

$P = 55 + 30 \times \frac{1}{3}$ Substitute (replace the letter with its value).

$P = 55 + 10$ Calculate carefully (use BIDMAS).

$P = 65$

The plumber is paid £65.

c   This time you are given the value of P.

$P = 55 + 30h$ Copy the formula.

$75 = 55 + 30h$ Substitute (replace the letter with its value).

This is an equation that you need to solve.

$20 = 30h$ Subtract 55 from both sides.

$\frac{20}{30} = h$ Divide both sides by 30.

$h = \frac{2}{3}$ hour

$\frac{2}{3}$ of 60 minutes = 40 minutes

The plumber needs to work 40 minutes to be paid £75.

Practise... 10.2 Substitution   

1   The area of a rectangle = length × width

Find the area of a rectangle with length = 6.2 cm and width = 5.1 cm.

2   A recipe book says a joint of meat needs to be cooked for 40 minutes per kg plus 20 minutes.

How long does it take to cook a joint of meat weighing 2.3 kg?

3   Use the formula $t = 3n + 8$ to find the value of t when n is 2.5

4   Janet is using the formula $v = u + at$ in Science.
She is told that $u = 12$, $a = 9.8$ and $t = 15$
Find the value of v.

5   Karim is working on polygons in mathematics. He has found the formula $T = 180(n - 2)$. He uses the formula to find the sum, T, of the interior angles of a polygon with n sides.

   a   Use Karim's formula to find the sum of the interior angles of a polygon with:

     i   4 sides   **ii**   9 sides.

   b   Use Karim's formula to find the number of sides of a polygon when T is:

     i   1440°   **ii**   720°.

6   Max is using the formula $F = 1.8C + 32$ to change temperatures from °C to °F.

   a   Use this formula to change the following from °C to °F.

     i $C = 10°C$ **ii** $C = 15°C$ **iii** $C = -3°C$

   b   Use this formula to change the following from °F to °C.

     i $F = 80°F$ **ii** $F = 52°F$ **iii** $F = 23°F$

   c   The midday temperature in Kingston is 35°C and the midday temperature in Cape Town is 98°F.

     Which city has the higher temperature?
     Give a reason for your answer.

7 Shaun hires a car for a day. The daily charge is £32 plus 10p per mile.

    a Write down a formula for the total charge when Shaun travels m miles in the car.

    b Use your formula to find the total cost of hiring the car when Shaun travels:

      i 256 miles   **ii** 755 miles.

    c Shaun has £40. What is the maximum number of miles he can do in the car?

8 Susan uses the formula $P = 2L + 2W$ to find the perimeter of rectangles with length L cm and width W cm.

    a Susan has a rectangle with $L = 5$ cm and $W = 4$ cm. Find P.

    b Susan has a rectangle with $P = 97$ cm. She measures L to be 28.4 cm. Work out W.

9 Fences are made up from identical lengths of wood as shown. They are available in different sizes, from size 1 upwards. Size 1 and size 2 fences are shown.

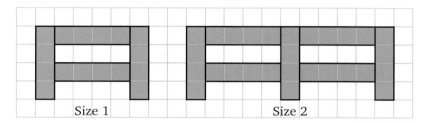

Size 1   Size 2

    a How many pieces of wood are needed to make a size 3 fence?

    b Write down a formula for the number of pieces of wood that are needed for a fence of any size.

    c Use your formula to find the number of pieces of wood required for a fence of size 8.

    d Use your formula to find the size of a fence containing 37 pieces of wood.

    e Jake says that his garden is very long so he needs a fence with 101 pieces of wood.
       Explain why Jake cannot be correct.

10 Dean is using the formula $H = \frac{1}{2}(50 - 0.1W)$

H is the height in centimetres of the bottom of the back of his car above the ground.

W is the weight in kilograms he puts in the back of his car.

    a Dean puts a weight of 90 kg in the back of his car.

       What is the value of H?

    b Dean takes the weight out of his car and three people get in.

       They weigh 60 kg, 80 kg, and 85 kg.

       What is the height of the bottom of the car above the ground?

    c For safe driving, the bottom of the back of Dean's car must be at least 10 cm above the ground.

       He has two people weighing 76 kg and 82 kg in the back of his car.

       He has a lawnmower weighing 30 kg in his car boot.

       How much extra weight could Dean put in his car boot and it still be safe to drive?

11 Mal is using the formula $s = ut + \frac{1}{2}at^2$ in science.

  a Find s when $u = 12$, $t = 5$, $a = -9.8$ **b** Find u when $s = 20$, $t = 3.9$, $a = -9.8$

12 May is using the formula $A = \pi r^2$ in mathematics to find the area of a circle with radius r cm.

Use the formula to find:

  a the area of a circle with radius 5 cm   **b** the radius of a circle with area 78 cm².

13 Terence is cooking a chicken in the oven.
The chicken weighs 7.2 pounds.
The cookbook says,

  'Cook in the oven for 20 minutes per
  pound plus an extra 20 minutes'.

  Take out of the oven and allow to rest
  for 10 minutes before carving.

Terence wants to carve the chicken at 12:45.

What is the latest time that Terence could put the chicken in the oven?

14 Gary the gasman is paid £46 call-out fee plus £27 for every hour he spends working on a repair.
Gary's diary for the week is nearly full. On Friday he can either do one large job likely to take
5 hours to complete, or two smaller jobs estimated to take a total of 4 hours to complete.

What option would earn Gary more money?

Give a reason for your answer.

10 Assess (k!)

1 Amy has n books on each of her six shelves.
Find a formula using words for the total number of books on Amy's shelves.

2 Gary takes a taxi to travel to the cinema one night. The fare is £5 plus £1 for every mile
travelled. The cinema is 7 miles from Gary's house.
How much does the taxi cost?

3 Use the formula $y = 2x + 6$ to find the value of y when x is 2.5

4 Eric the electrician charges a £38 call-out fee plus £24 for each hour he spends working.

  a Write a formula for the cost, £C, of calling Eric out for a job that takes h hours.

  b Use your formula to find the cost of a job which takes Eric:

    i 3 hours 30 minutes   **ii** 2 hours 45 minutes   **iii** 40 minutes.

5 Jill is using the formula $F = 1.8C + 32$ to convert temperatures.
She notices that a box of ice cubes says 'store at $-18°C$ ($0°F$)'.
Use the formula to check that $-18°C$ is the same as $0°F$. Show your working.

6 Faisal is using the formula $V = 3S + 5W$

  a Find V when $S = 2.5$ and $W = 3\frac{1}{4}$ **b** Find W when $S = 6$ and $V = 30$

7 Use the formula $C = \dfrac{(A + 1)D}{9}$ to find C when $A = 11$ and $D = 5$

8 This shape is made from two rectangles.
Find formulae for:

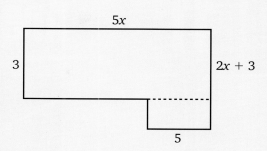

a the perimeter

b the area.

Write your formulae in their simplest forms.

9 Electric cable weighs 2.1 kg per metre. It is sold in containers which weigh 3.5 kg.

a What is the weight of 3 metres of electric cable?

b Write down a formula for the weight of x metres of electric cable including the container.

c John's car can carry a total of 50 kg on its roof. What is the maximum length of electric cable he can put on his car roof?

10 Jack's wood burner burns logs at a rate of 2.5 kg per hour. He uses his wood burner for 5 hours each day in winter.

a What is the amount of wood Jack burns in:

i 1 day   **ii** 2 days   **iii** n days?

> **Hint**
> 1 tonne = 1000 kg

b Jack buys 3 tonnes of logs. He pays £90 per tonne plus £20 delivery.

i Will this wood last him all through the three months of winter?

ii How much per day will it cost Jack to burn this wood in his burner?

11 Use the following formula

$$\text{speed} = \frac{\text{distance}}{\text{time}}$$

to help you in this question:

A cyclist travels for 2 hours at x mph. The cyclist then cycles at 1 mph less for the next 3 hours. The cyclist rides 42 miles in total. Find the value of x.

AQA Examination-style questions

1 After exercise you can work out your fitness index, F.
You need to know:
    your exercise time in seconds (T)
    the number of your pulse beats in three 30-second intervals after you have stopped exercising (a, b and c).

Tony is working out his fitness index.

a i Tony exercises for 3 minutes 30 seconds.
Work out T.   *(1 mark)*

ii After exercise he obtains $a = 70$, $b = 55$ and $c = 45$
Work out $a + b + c$.   *(1 mark)*

iii Work out F, Terry's fitness index, using the formula:

$$F = \frac{50T}{a + b + c}$$
        (2 marks)

b Your fitness grade can be worked out from your fitness index, F, using this table.

Fitness index, F	less than 50	50 to 59	60 to 69	70 to 79	80 to 89	$\geqslant 90$
Fitness grade	Very poor	Poor	Fair	Good	Excellent	Superb

What is Terry's fitness grade?   *(1 mark)*

AQA 2009

Area and volume

Did you know?

Optical prisms

An optical prism is transparent with flat, polished surfaces that separate a beam of white light into its spectrum of colours.

Before Isaac Newton's experiments, it was thought that white light was colourless, and it was the prism producing the colour. Newton did experiments that made him think that all the colours were already in light, and that particles of light were spread out because different coloured particles went through the prism at different speeds, making a rainbow-like effect.

The traditional geometrical shape is that of a triangular prism with a triangular base and rectangular sides, but there are other shapes that are known as prisms.

Key terms

dimension
solid
cross-section
prism
cube
cuboid
volume
surface area
face
net

You should already know:

✔ how to find the area of rectangles, triangles, parallelograms and circles

✔ how to convert between metric units, e.g. centimetres to metres

✔ how to draw nets of solids.

Learn... 11.1 Volume of a cuboid

Shapes such as rectangles have two **dimensions**: length and width.

Shapes that have a third dimension such as thickness or height are called **solids**.

Solids which have the same **cross-section** all the way through the shape are called **prisms**.

Two of the most commonly known prisms are **cube** and **cuboid**.

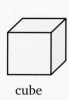

cube
(cross-section
is a square)

cuboid
(cross-section
is a square)

The **volume** of a shape is the amount of space it occupies.
It is measured in cubic units.

For example, this cube is 1 cm long, 1 cm wide and 1 cm high.

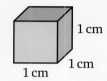

1 cm

1 cm

1 cm

The amount of space it takes up is $1 \text{ cm} \times 1 \text{ cm} \times 1 \text{ cm}$, which is 1 cubic centimetre, written 1 cm^3.

Volume can be measured by counting the number of 1 centimetre cubes in a three-dimensional shape.

Counting the number of 1 centimetre cubes in this shape gives a volume of 16 cubes or 16 cm^3.

Counting cubes can take a long time.

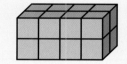

In this cuboid each layer has $4 \times 2 = 8$ cubes (the length $\times$ the width of the cuboid), and there are 2 layers (the height of the cuboid).

The total number of cubes is $4 \times 2 \times 2 = 16$

So

    volume of a cuboid = length $\times$ width $\times$ height

As the length $\times$ width gives the area of the base, the formula can also be written:

    volume = area of base $\times$ height

Units of volume and area

Volume is always measured in cubic units such as cubic millimetres (written mm^3), cubic centimetres (written cm^3) or cubic metres (written m^3).

Sometimes it is necessary to convert between these units.

There are 100 centimetres in 1 metre.

How many cubic centimetres (cm^3) are there in 1 cubic metre (m^3)?

This cube has sides of 1 metre. Its volume is $1 \text{ m} \times 1 \text{ m} \times 1 \text{ m} = 1 \text{ m}^3$

If the dimensions of the cube were given in centimetres then each side would measure 100 cm.

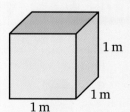

1 m

1 m

1 m

The volume would be $100 \text{ cm} \times 100 \text{ cm} \times 100 \text{ cm} = 1\,000\,000 \text{ cm}^3$

So $1 \text{ m}^3 = 1\,000\,000 \text{ cm}^3$

In the same way the area of the base can be found in square metres or square centimetres.

The area of the base of this cube is $1 \text{ m} \times 1 \text{ m} = 1 \text{ m}^2$

Working in centimetres the area of the base is $100 \text{ cm} \times 100 \text{ cm} = 10\,000 \text{ cm}^2$

So $1 \text{ m}^2 = 10\,000 \text{ cm}^2$

Example: Work out the volume of this cuboid.

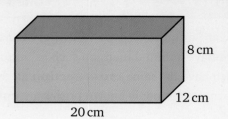

Solution: Using the formula, volume = length × width × height

$$= 20 \text{ cm} \times 12 \text{ cm} \times 8 \text{ cm}$$

$$\text{Volume} = 1920 \text{ cm}^3$$

Example: A cuboid has a volume of 260 m³.
The area of the base of the cuboid is 65 m².
Work out the height of the cuboid.

Solution: Using the formula,   volume = length × width × height
and the formula,   area of base = length × width
gives: volume = area of base × height

$$260 = 65 \times \text{height}$$
$$\frac{260}{65} = \text{height} \qquad \text{Divide both sides by 65.}$$
$$\text{Height} = 4 \text{ m}$$

Example:
a   Convert 24 500 cm² to square metres.

b   Convert 5 m² to square centimetres.

c   Convert 7 250 000 cm³ to cubic metres.

Solution: 1 m² = 1 m × 1 m = 100 cm × 100 cm = 10 000 cm²

a   To convert cm² to m² divide by 10 000.
24 500 ÷ 10 000 = 2.45 m²

b   To convert m² to cm² multiply by 10 000.
5 × 10 000 = 50 000 cm²

c   1 m³ = 1 m × 1 m × 1 m = 100 cm × 100 cm × 100 cm = 1 000 000 cm³
To convert cm³ to m³ divide by 1 000 000.
7 250 000 ÷ 1 000 000 = 7.25 m³

Practise...   11.1   Volume of a cuboid   G F E D C

G

1   These solid shapes are made from 1 cm cubes.
Find the volume of each shape.

a   **b**   **c**   **d**

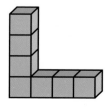
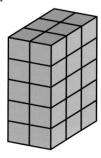
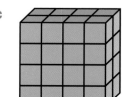
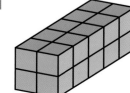

2 The table gives the measurements of some cuboids.

Copy and complete the table.

Cuboid	Length (cm)	Width (cm)	Height (cm)	Volume (cm³)
a	6	5	3	
b	15	8	7	
c	9.6	4.8	5	
d	42.2	25	4.5	

3 Find the volume of each of the following cuboids.

State the units of each answer.

a

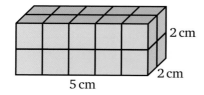
5 cm, 2 cm, 2 cm

c

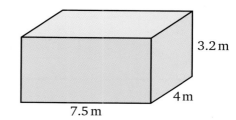
3.2 m, 4 m, 7.5 m

b

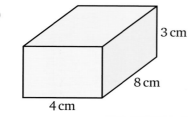
3 cm, 8 cm, 4 cm

d

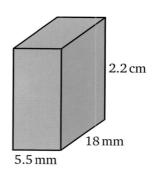
2.2 cm, 18 mm, 5.5 mm

Hint
Remember to change all lengths to the same units.

4 Which has the larger volume: a cuboid measuring 5 cm by 7 cm by 3 cm, or a cube of side 5 cm?

Show your working.

5 **a** Convert the following areas to square centimetres.

   i 4.6 m²   **ii** 23 m²   **iii** 9 m²   **iv** 0.5 m²

  b Convert the following areas to square metres.

   i 300 000 cm²   **ii** 75 000 cm²   **iii** 57 600 cm²   **iv** 8 500 cm²

6 Find the volume of these solids.

a

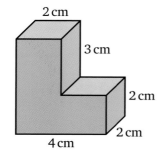
2 cm, 3 cm, 2 cm, 2 cm, 4 cm

b

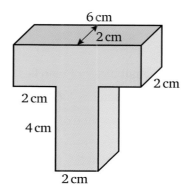
6 cm, 2 cm, 2 cm, 2 cm, 4 cm, 2 cm

Hint
Divide each solid into cuboids.

D

7 This cuboid has a volume of 432 m³.

Work out the height, h, of the cuboid.

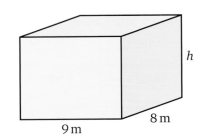

Not drawn accurately

8 The table gives the measurements of some cuboids.

Copy and complete the table.

Cuboid	Length (cm)	Width (cm)	Height (cm)	Volume (cm³)
a	7	11		385
b	14		19	7182
c		4.1	10	282.9
d	15.5		3.8	565.44

C

9 Convert the following volumes to cubic centimetres.

a 8 m³   b 3.2 m³   c 0.765 m³   d 0.0568 m³

⚠10 A 25 mm square hole is cut right through the centre of a cuboid as shown.

Find the volume of the remaining cuboid.

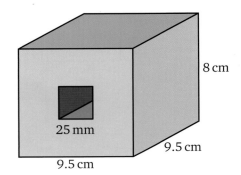

⚙11 A concrete beam is 20 metres long, 0.3 metres wide and 0.4 metres thick.

How many cubic metres of concrete are needed to make the beam?

⚙12 A garden is a rectangle of length 9 m and width 8.4 m.

The garden is to be covered with topsoil to a depth of 5 cm.

Topsoil costs £82.55 for a bag containing 1 m³ of topsoil or £38 for a bag containing 0.4 m³.

Work out the cheapest cost of the topsoil.

?13 Twelve small boxes of matches are to be packed tightly into a carton.

Each box of matches has length 5 cm, width 3.5 cm and height 1.5 cm.

Work out the volume of the carton.

Work out the possible dimensions of the carton.
(The boxes of matches can be packed in layers.)

14 Do these cuboids have the same volume?

Show your working.

a
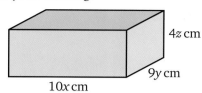
4z cm
9y cm
10x cm

b
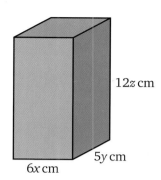
12z cm
5y cm
6x cm

Learn... 11.2 Volume of a prism

A cuboid is a prism with a rectangular cross-section.

Remember that a prism is any solid with the same cross-section all the way through.

Here are some prisms with their cross-sections shaded.

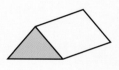
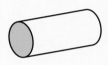

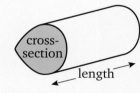

cross-section
length

AQA Examiner's tip

The formula for the volume of a prism will be given to you on the second page of the examination paper.

volume of prism = area of cross-section × length (or height)

This formula applies to any prism.

Example: Work out the volume of each of these prisms.

a
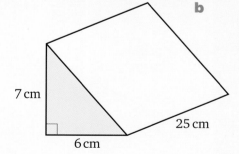
7 cm
6 cm
25 cm

b
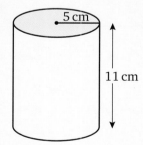
5 cm
11 cm

Solution: **a** The cross-section of this prism is a right-angled triangle with base 6 cm and perpendicular height 7 cm.

Using the formula, area of triangular cross-section = $\frac{1}{2}$ × base × height

$= \frac{1}{2}$ × 6 × 7

Area of triangular cross-section = 21 cm²

AQA Examiner's tip

Remember:
- area is measured in square units
- volume is measured in cubic units.

Using the formula, volume = area of cross-section × length

= 21 × 25

Volume = 525 cm³

b The cross-section of this prism is a circle of radius 5 cm.

Using the formula, area of a circle $= \pi r^2$

$$= \pi \times 5^2$$

Area of cross-section $= 25\pi$

Using the formula, volume $=$ area of cross-section $\times$ height

$$= 25 \times \pi \times 11$$

Volume $= 863.9$ cm^3 (to 1 d.p.)

AQA *Examiner's tip*

Shade in the cross-section to make it clear what area you are working out.

Practise... 11.2 Volume of a prism 🄺 G F E D C

C

1 The area of the cross-section of each prism is given in these diagrams.

Work out the volume of each prism.

Remember to state the units of each answer.

a

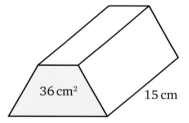

22 cm² 8 cm

c

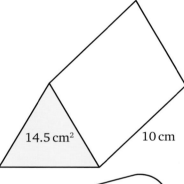

14.5 cm² 10 cm

b

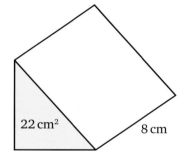

36 cm² 15 cm

d

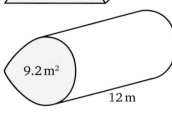

9.2 m² 12 m

2 Work out the volume of each of these **triangular prisms**.

a

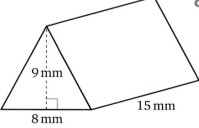

9 mm 15 mm 8 mm

c

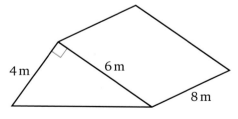

4 m 6 m 8 m

b

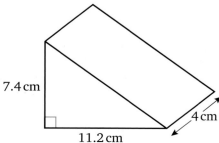

7.4 cm 11.2 cm 4 cm

3 Work out the volume of each **cylinder**.

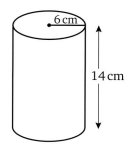

a

6 cm

14 cm

b

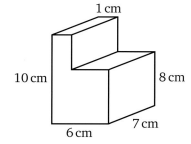

8.4 m

32 m

c

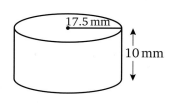

17.5 mm

10 mm

4 A prism has a volume of 132 cm³. The area of the cross-section of the prism is 33 cm².

Work out the height of the prism.

Hint
Draw a sketch of the cross-section to help you.

5 For each prism shown below, work out:

i the area of the cross-section **ii** the volume.

a

1 cm

10 cm

8 cm

7 cm

6 cm

c

3 cm

4 cm

17.5 cm

8 cm

b

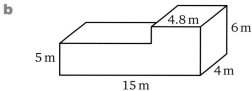

4.8 m

6 m

5 m

4 m

15 m

6 A cylinder of height 3.2 m has a volume of 15.68 m³.

Work out the area of the base of the cylinder.

△ 7 The diagram shows a plastic pipe of internal radius 4 cm and length 60 cm.

The plastic has a thickness of one centimetre.

Calculate the volume of plastic in the pipe.

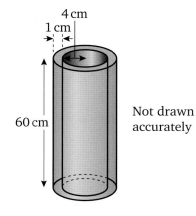

4 cm

1 cm

60 cm

Not drawn accurately

⚙ 8 At a pre-school playgroup, each of the 36 children is given a beaker of milk.
The beakers are cylinders of radius 3 cm and height 8 cm and are three-quarters full.

Each milk carton contains 2.2 litres of milk.

Susie says that three cartons will be enough for all the children.

Is she correct?

Show your working.

Hint
1000 cm³ = 1 litre

9 The volume of a prism is 90 cm³.
Find three different shapes of prism with this volume.
Give the dimensions of each one.

10 These two prisms have the same volume. Work out h.

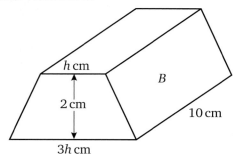

Not drawn
accurately

Learn... 11.3 Surface area of a prism 🔑

The total **surface area** of a three-dimensional shape is the sum of the area of all the **faces** of the shape.

For example, a cube where the length of each side is 3 cm is made up of six square faces, each measuring 3 cm by 3 cm.
The area of each face is 3 cm × 3 cm = 9 cm²
The total surface area of all six faces is 6 × 9 cm² = 54 cm²

It is often useful to draw the **net** of a solid to help you to see the individual areas.

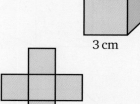

AQA Examiner's tip

Remember to give the correct units for your answer. Area is always measured in square units.

Example: Work out the surface area of this triangular prism.

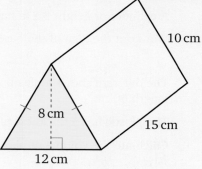

Solution: Sketch the net.

The middle rectangle (A) has area
12 × 15 = 180 cm²

The two outer rectangles (B and C)
each have area 10 × 15 = 150 cm²

The triangles (D and E) each have
area $\frac{1}{2}$ × 12 × 8 = 48 cm²

Total surface area
= 180 + (2 × 150) + (2 × 48)
= 576 cm²

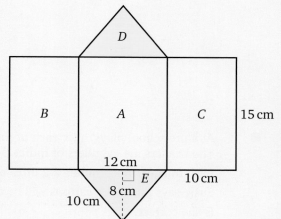

Example: Work out the area of the curved surface of this cylinder.

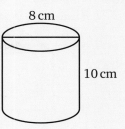

Solution: You are asked to work out the **curved** surface.
This means you don't have to include the flat ends.

The curved surface of the cylinder forms a rectangle when the net is drawn.

The length of the rectangle is equal to the circumference of the circular end of the cylinder.

Circumference = $2\pi r$ so the area of the curved surface = $2\pi r \times h$

In the cylinder shown, the radius is 4 cm (half the diameter).

area of the curved surface = $2 \times \pi \times 4 \times 10 = 80\pi$
or 251.3 cm² (to 1 d.p.)

If you are asked to find the area of the **whole** cylinder, you must
add the area of the circular top and base to the area of the curved surface.

Practise...   11.3 Surface area of a prism 🔑   G F E D C

1   Here is a net of a cuboid.
Work out the surface area of the cuboid.

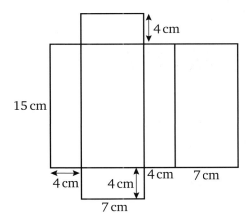

D

2   Here is a net of a triangular prism.
Work out the surface area.
Give your answer in:

a   m²   **b**   cm²

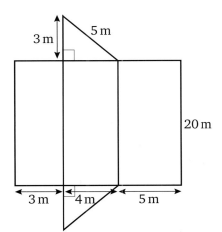

3   Calculate the total surface area of cubes with these side lengths.

a   7 cm   **b**   10 cm   **c**   5.4 cm

C

C

4   **a**   Calculate the total surface area of these cuboids.

    i

       3 cm

       8 cm

       4 cm

    ii

       22 mm

       18 mm

       5.5 mm

    b   Calculate the total surface area of these triangular prisms.

      i

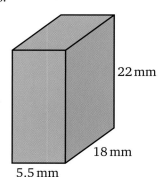

       9 mm   15 mm

       8 mm

      ii

       7.4 cm   4 cm

       11.2 cm

> **Bump up your grade**
>
> For a Grade C make sure you know how to work out the surface area of a solid.

5   Which of these units are correct for the surface area of a solid?

    mm^2 cm^3 m^2 mm km^2 cm^2

6   Calculate the area of the curved surface of each of these cylinders.

    a   6 cm   **b**   **c**   17.5 mm

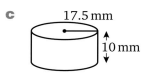

       14 cm   8.4 m   ←32 m→   10 mm

7   Jack says that the formula for the surface area of a cuboid is length × width × height.

    Is he correct?

    Give a reason for your answer.

⚠ 8   Here are two closed cylinders. (They have a top and a base.)

    Calculate the total surface area of each cylinder.

    a   6 cm   **b**

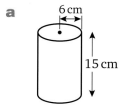 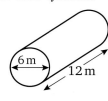

       15 cm   6 m   12 m

⚙ 9   Andrew makes wooden jewellery boxes in the shape of cuboids of length 20 cm, width 15 cm and depth 10 cm. Each box has a lid.

    The wood costs £18.75 for 1 square metre and on average he wastes 10% of each square metre.

    He varnishes the outside of each box when he has made them.

    A tin of varnish will cover an area of 1.5 m² and costs £7.99.

    What is the cost of making each jewellery box?

> **Hint**
>
> Draw a diagram to help you.

10 In this cuboid the surface areas of the three faces shown are 15 cm², 12 cm² and 20 cm².

Work out the volume of the cuboid.

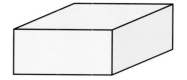

Not drawn accurately

11 Natalie has three wooden cubes that she is using in DT. The smallest cube has sides of length 3 cm. The medium-sized cube has sides of length 6 cm. The largest cube has sides of length 10 cm. She sticks them together to make the solid shown.

Natalie wants to paint the solid red. The tin of red paint she uses will cover an area of 0.5 m².

Will she have enough paint? Show your working.

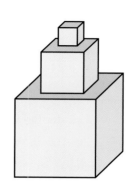

Not drawn accurately

12 The area of the curved surface of this cylinder is equal to three times the area of both ends added together. Express h in terms of r.

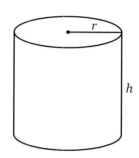

Not drawn accurately

Assess (k!)

1 These solids are made from cubes of side 1 cm.

Find the volume of each solid.

a

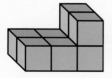

Not drawn accurately

b

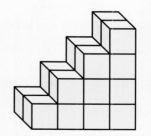

Not drawn accurately

2 These solids are made from cubes of side 1 m. Find the volume of each solid.

a

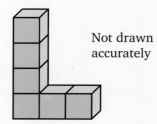

Not drawn accurately

b

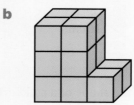

Not drawn accurately

G

3   A prism has a total surface area of 0.75 m².

What is the surface area of the prism in cm²?

4   The table shows the measurements of some cuboids.

Cuboid	Length	Width	Height
a	14 cm	6 cm	3 cm
b	45 mm	22 mm	10 mm
c	3.2 m	6 m	4 m
d	12.4 cm	15.5 cm	11 cm

i   Work out the volume of each cuboid.

ii   Calculate the total surface area of each cuboid.

Remember to state the units of each answer.

5   Calculate the volume of this cuboid.

Give your answer **a** in cm³ **b** in mm³.

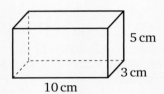

6   Calculate the volume of each of these prisms.

a

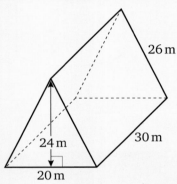

b

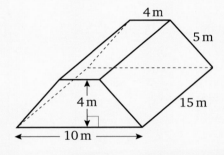

7   Here is a list of six formulae.

πr^2	length × width × height	2 × length + 2 × width
area of cross-section × length	$2\pi r$	$\frac{1}{2}$ × base × height

Write down which of these formulae represent:

a   a length   **b**   an area   **c**   a volume.

8   Work out the total surface area of this triangular prism.

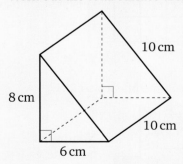

9   A metal pole is in the shape of a solid cylinder. It has a radius of 1.5 m and a length of 17 m.

Work out the volume of metal used for the pole.

AQA Examination-style questions

1 Centimetre cubes are fitted together to make a solid as shown on the left.

The solid is packed into a box as shown on the right.

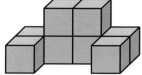 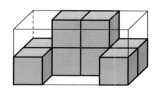

The box is a cuboid.
Work out the volume of the box.

(3 marks)
AQA 2009

2 A cuboid has a volume of 75 cm³.

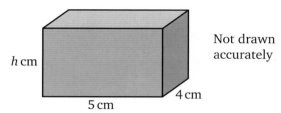

h cm

Not drawn accurately

4 cm

5 cm

The length is 5 cm.
The width is 4 cm.
Find the height, *h* cm.

(2 marks)
AQA 2008

3 The diagram shows a block of wood with uniform cross-section.
The cross-section is made of rectangles.
The block is 65 cm long.

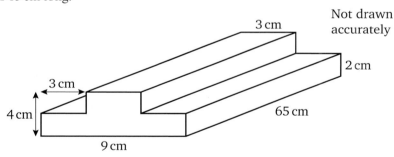

3 cm   Not drawn accurately

2 cm

3 cm

4 cm   65 cm

9 cm

Calculate the volume of the block.
State the units of your answer.

(5 marks)
AQA 2009

4 The diameter of a cylindrical water-butt is 55 cm.
The height is 82 cm.

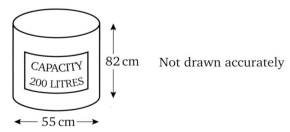

CAPACITY
200 LITRES

82 cm   Not drawn accurately

← 55 cm →

One litre = 1000 cm³
The label states that the capacity of the water-butt is 200 litres.
Is this accurate?
You **must** show your working.

(4 marks)
AQA 2008

12 Measures

Objectives

Examiners would normally expect students who get these grades to be able to:

G

decide which is the most appropriate unit of measurement to use in everyday situations

convert between metric units

measure the length of a line

F

convert between metric and imperial units such as kilograms and pounds

measure and scale a line

make sensible estimates of lengths

E

convert between metric and imperial units such as speed, for example, convert 80 km/h to mph

D

calculate average speeds

C

use compound measures, such as speed

recognise that measurements to the nearest unit may be inaccurate by up to one half unit in either direction.

Did you know?

Going metric

We didn't always pay for things using pounds and pence. Before 1971 we used pounds, shillings and pence.

We didn't always measure lengths using metres. Before 1965 we used 'imperial units'. We measured lengths using yards, feet and inches. Some people still do!

We still use some very old measures. For example, the length of a cricket pitch is 22 yards.

Key terms

unit   conversion factor

mass   compound measure

capacity

You should already know:

✔ how to multiply and divide by 10, 100 and 1000.

Learn... 12.1 Measurements and reading scales

Measurements

Whenever you measure something you only get an approximate answer.

If you measure the length of the school hall you will probably give your answer to the nearest metre.

If you wanted to find a more accurate answer you might measure to the nearest centimetre.

When you measure something, you have to make these decisions.

- Decide on the most appropriate **units** to use.
- Decide what to use to make the measurement.
- Read the measurement and decide how to record it.

Units of **length** are:

kilometres (km), metres (m), centimetres (cm) and millimetres (mm).

Units of **mass** are:

tonnes (t), kilograms (kg), grams (g) and milligrams (mg).

Units of **capacity** are:

litres (l) and millilitres (ml).

Units of **time** are:

hours (h), minutes (min) and seconds (s).

> AQA **Examiner's tip**
>
> **All** metric units are linked by multiples of 10, but units of time are not.
>
> 60 seconds = 1 minute
>
> 60 minutes = 1 hour.

Example: Which of the units kilometres (km), metres (m), centimetres (cm) and millimetres (mm) would be used to measure:

a the distance from York to London

b the length of a pen

c the length of a cricket pitch

d the diameter of a pen?

Solution: a Kilometres. As it is a long way from York to London you would use large units.

b Centimetres. As the length of a pen is much less than a metre, it would be sensible to use centimetres. Millimetres could be used, but most pens would be more than 100 millimetres long.

c Metres. The length of a cricket pitch is between 1 m and 1 km, so it is best to use metres.

d Millimetres. The width of most pens is between 1 mm and 1 cm, so millimetres are the best units to use.

Reading scales

When you measure anything you need to read from a scale. Only the main numbers are written on a scale. The gap between these numbers is divided up so that you can read between the main numbers.

Example: a How much water is in the measuring cylinder?

b   What weight do the weighing scales show?

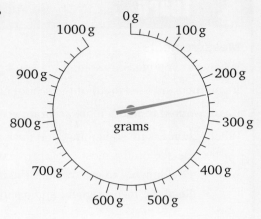

c   What is the length of the nail?

mm	0	10	20	30	40	50	60	70	80	90	100	110	120	130	140	150
cm	0	1	2	3	4	5	6	7	8	9	10	11	12	13	14	15

Solution: **a**   There are 5 divisions between 50 ml and 100 ml. That is 5 divisions for 50 ml. Each division is 50 ml ÷ 5 = 10 ml. The level of the water is one division above 50 ml. The measuring jug contains 60 ml of water.

b   There are 5 divisions between 200 g and 300 g. That is 5 divisions for 100 g. Each division is 100 g ÷ 5 = 20 g. The pointer is at 2 divisions more than 200 g.
200 g + 40 g = 240 g. The scales show 240 g.

c   The ruler is marked in millimetres (mm) and centimetres (cm). We can give the length of the nail in different ways.

To the nearest centimetre:
The nail is between 4 cm and 5 cm long.
The point of the nail is nearer 5 cm than 4 cm.
The nail is 5 cm to the nearest centimetre.

To the nearest millimetre:
The nail is between 48 mm and 49 mm.
It is nearer to 49 mm.
The nail is 49 mm to the nearest mm.

Using centimetres and millimetres:
1 cm is 10 mm and 1 mm is 0.1 cm.
4 cm is 40 mm and 9 mm is 0.9 cm.
49 mm can be written as 4 cm and 9 mm, or 4.9 cm.

Practise...   12.1 Measurements and reading scales

G F E D C

1   Measure each of the following lines to the nearest centimetre.

a _____

b _____

c _____

d _____

2   The diagram shows a pen and a ruler.

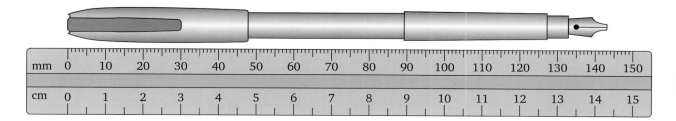

G

a   What is the length of the pen to the nearest centimetre?

b   What is the length of the pen to the nearest millimetre?

3   **kilometres, metres, centimetres, millimetres**

Which of these units would be most appropriate for measuring each of the following?

a   The height of Blackpool tower

b   The thickness of a paperback book

c   The length of a football pitch

d   The thickness of a matchstick

e   The distance from Leeds to Manchester

f   The height of a chair seat above the ground

g   The thickness of a coin

h   The height of a building

i   The distance from New York to London

j   The height of a tree

4   **tonnes, kilograms, grams, milligrams**

Which of these units would be most appropriate for measuring each of the following?

a   The weight of a car

b   The weight of a bag of sugar

c   The weight of an adult

d   The weight of a sugar lump

e   The weight of the amount of butter in a cake

f   The weight of a bag of sweets

g   The weight of a feather

5   **litres, millilitres**

Which of these units would be the most appropriate for measuring the capacity of each of the following?

a   A cup

b   A reservoir

c   A bath

d   A bucket

e   A car's petrol tank

f   A teaspoon

G

6 Jim is baking a cake. He weighs butter, flour and sugar.

How much of each does he use?

a Butter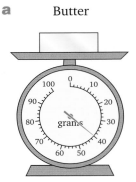

b Flour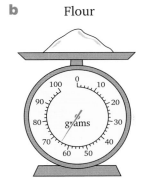

c Sugar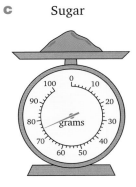

7 Gina is riding her bike. The speedometers show her speed in km/h.

What speeds do they show?

AQA *Examiner's tip*

Remember to work out how much each division of the scale is worth.

8 These thermometers show the temperature in different rooms in a house.

Write down the temperature in each room.

a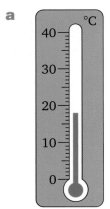

Kitchen

b

Bathroom

c

Bedroom

d

Dining room

G
F

9 **a** Estimate the length and width of this textbook.

b Check your estimates by measuring.

10 **a** Estimate the length and width of your classroom.

b Check your answers by measuring. (Agree with your teacher when would be a convenient time to measure the classroom.)

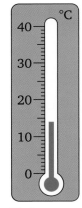

F

11   How many millilitres of milk are in this jug?

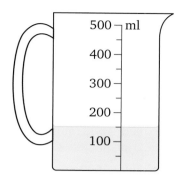

Learn... 12.2 Conversion between metric units

To convert from one metric unit to another, multiply or divide by a **conversion factor**.

To convert from a large unit to a smaller unit, multiply by the conversion factor as there will be more of the smaller units.

To convert from a small unit to a larger unit, divide by the conversion factor as there will be fewer of the larger units.

You are expected to know the following conversions between metric units.

Distance

1 kilometre = 1000 metres	(1 km = 1000 m)
1 metre = 1000 millimetres	(1 m = 1000 mm)
1 metre = 100 centimetres	(1 m = 100 cm)
1 centimetre = 10 millimetres	(1 cm = 10 mm)

Mass

1 tonne = 1000 kilograms	(1 t = 1000 kg)
1 kilogram = 1000 grams	(1 kg = 1000 g)
1 gram = 1000 milligrams	(1 g = 1000 mg)

> **AQA** *Examiner's tip*
>
> You are expected to know the metric equivalents as they are listed here.

Capacity

1 litre = 1000 millilitres	(1 l = 1000 ml)
1 litre = 100 centilitres	(1 l = 100 cl)
1 centilitre = 10 millilitres	(1 cl = 10 ml)
1 litre = 1000 cubic centimetres	(1 l = 1000 cm³)

The metric system is based on 1000s.

'kilo' means 1000, so 1 'kilo'metre means 1000 metres.

'milli' means $\frac{1}{1000}$, so 1 'milli'metre means $\frac{1}{1000}$ of a metre.

'centi' means $\frac{1}{100}$ so 1 'centi'metre means $\frac{1}{100}$ of a metre.

Example: Convert the following lengths to metres.

     a   14 km

     b   23.5 cm

     c   45 mm

Solution: **a**   There are 1000 metres in every 1 kilometre, so the conversion factor is 1000.

To change 14 kilometres to metres you multiply by 1000. (This is because you are converting from large units to small units so there will be more of them.)

Using the fact that 1 kilometre = 1000 metres

$$14 \text{ kilometres} = 14 \times 1000 \text{ metres}$$
$$= 14\,000 \text{ m}$$

     b   1 metre is made up of 100 centimetres, so the conversion factor is 100.

To change 23.5 cm to metres you divide by 100. (This is because you are converting from small units to large units so there will be fewer of them).

$$23.5 \text{ cm} = 23.5 \div 100 = 0.235 \text{ m}$$

     c   There are 1000 millimetres in every metre, so the conversion factor is 1000.

To change 45 mm to metres you divide by 1000. (This is because you are converting from small units to large units so there will be fewer of them).

$$45 \text{ mm} = 45 \div 1000 = 0.045 \text{ m}$$

or using the fact that 1 metre = 1000 millimetres

$$1000 \text{ millimetres} = 1 \text{ metre}$$
$$1 \text{ millimetre} = \tfrac{1}{1000} \text{ metre} = 0.001 \text{ metres}$$
$$45 \text{ millimetre} = 45 \times 0.001 \text{ metres}$$
$$= 0.045 \text{ m}$$

> **AQA** *Examiner's tip*
>
> Remember to give the correct units with your answer.

Example: Gerald's car has a mass of 1.4 tonnes. Convert this to kilograms.

Solution: There are 1000 kilograms in 1 tonne, so the conversion factor is 1000.

To convert from 1.4 tonnes to kilograms you multiply by 1000 (you are converting from large units to small units so there will be more of them).

$$1.4 \text{ tonnes} = 1.4 \times 1000 \text{ kg}$$
$$= 1400 \text{ kg}$$

Practise...

12.2 Conversion between metric units

G F E D C

G

1   Measure and write down the lengths of these lines in millimetres.

     a ————————————————

     b ——————————

     c —————————————————

     d ————

     e ——

2   What is the length of each line in Question 1 in centimetres?

3 Jemima measured the distance from her house to school as 3 kilometres.
What is this distance in metres?

4 Gill measured the length of her classroom as 4.5 metres.
What is this in centimetres?

5 Harold walked 1250 metres across the school playing field.
How far did Harold walk in kilometres?

6 Convert the following to kilograms.

    a 1.2 tonnes   **c** 0.9 tonnes   **e** 2500 g

    b 3.1 tonnes   **d** 2000 g   **f** 250 g

7 Samantha buys a bottle of juice containing 70 centilitres.

    a How many millilitres does the bottle contain?

    b How many litres does the bottle contain?

8 Copy and complete the following, filling in the missing units.

    a 2 km = 2000 …   **f** 30 cl = 300 …

    b 20 cm = 200 …   **g** 1.5 kg = 1500 …

    c 30 mm = 3 …   **h** 6.2 m = 6200 …

    d 400 m = 0.4 …   **i** 6.2 m = 620 …

    e 3 t = 3000 …   **j** 1500 ml = 1.5 …

9 Jeremy buys a bag of flour weighing 2.5 kg. He bakes 10 cakes, each of which uses 65 g of flour.
How much flour does he have left?

10 Tom walks 950 metres to school every morning. He walks the same distance home every afternoon.
How many kilometres does Tom walk in one week?

11 Fay has a recipe for fruit punch using 1.5 litres of orange juice, 2 litres of apple juice, 1 litre of water and 2.5 litres of lemonade. She makes some fruit punch using this recipe, but only needs half the quantity.

    a How much does she need of each ingredient? (Give your answers in millilitres.)

    b She shares the punch equally with six of her friends.
How many 250 ml glasses will they each have?

12 Jan bought a 2 m roll of wrapping paper for presents. She used 45 cm for James's present, and 97 cm for Carlos's present. She needs 65 cm to wrap Ian's present.
Does she have enough paper or will she need another roll? Show your working.

13 **a** Janet has four pieces of wood. They are 2 cm, 3 cm, 6 cm and 12 cm long.

        i Show how these can be used to measure lengths of 1 cm, 4 cm, and 7 cm.

        ii What other lengths can be measured with these four pieces of wood?

    b You have five pieces of wood.
They can be any length you want.
What lengths would you choose in order to measure the largest possible number of lengths?

> **Hint**
> Work systematically to find all the possibilities.

 **Learn...**

12.3 Conversion between metric and imperial units

To convert between metric and imperial units, you multiply or divide by a conversion factor.

You need to know, and be able to use, the following conversion factors.

5 miles ≈ 8 kilometres
4.5 litres ≈ 1 gallon
2.2 pounds ≈ 1 kilogram
1 inch ≈ 2.5 centimetres

> **Hint**
>
> The symbol ≈ means approximately equals. So that a distance of 5 miles is approximately equal to 8 kilometres.

> **AQA Examiner's tip**
>
> These are the only equivalents you need to know. Learn them, as they may not be given in the exam. Other conversions may be given.

The conversion factors give you information about which is the smaller unit if you are not sure. For example, 2.2 pounds ≈ 1 kilogram, so pounds must be smaller than kilograms.

For conversions between miles and kilometres, 5 miles ≈ 8 km

This can be difficult to use in practice.

You can use the conversion factor for 1 km: $\frac{5}{8}$ miles ≈ 1 km

so 0.625 miles ≈ 1 km

or 1 mile ≈ $\frac{8}{5}$ km

so 1 mile ≈ 1.6 km

> **AQA Examiner's tip**
>
> 0.63 miles ≈ 1 km is used in exam papers.

All of these give exactly the same answers, but some may be easier to use than others at times.

> **AQA Examiner's tip**
>
> Use 'multipliers' in ratio to convert. Remember there are **more** kilometres than miles for the same distance. For example:
>
> 30 miles = 30 × $\frac{8}{5}$ km = 48 km (greater answer)
>
> 120 km = 120 × $\frac{5}{8}$ miles = 75 miles (fewer miles)

Example: Convert 42 miles to kilometres.

Solution: 1.6 km ≈ 1 mile, so the conversion factor is 1.6

You are converting from large units to smaller units. (You know this because 1 mile is more than 1 km). So you multiply by 1.6 (as you are converting from a larger unit to a smaller unit, there will be more of them).

42 miles ≈ 42 × 1.6 kilometres

≈ 67.2 kilometres

≈ 67 kilometres (nearest km)

Example: Convert 50 litres to gallons.

Solution: 4.5 litres ≈ 1 gallon, so the conversion factor is 4.5

Litres are smaller than gallons. (You can tell this because 1 gallon is more than 1 litre; in fact it is about 4.5 litres.) So you divide by 4.5 (as you are converting from a smaller unit to a larger unit, there will be fewer of them).

50 litres ≈ 50 ÷ 4.5 gallons

50 litres ≈ 11.1 gallons

or using the fact that 4.5 litres ≈ 1 gallon

Therefore 1 litre ≈ $\frac{1}{4.5}$ gallons

50 litres ≈ 50 × $\frac{1}{4.5}$ gallons

50 litres ≈ 50 ÷ 4.5 ≈ 11.$\dot{1}$ gallons

50 litres ≈ 11 gallons

Practise...

12.3 Conversion betwen metric and imperial units

G F E D C

F

1 Convert 3 gallons to litres.

2 Convert 3 pounds to kilograms.

3 Alan measures a rod in a science lesson. It measures 6 inches.
What is this in centimetres?

4 A recipe for jam uses 10 pounds of sugar.
Approximately how many kilograms is this?

5 Chicken is priced at £8.50 per kilogram.
Approximately how much is this per pound?

6 Petrol is advertised at £1.12 per litre.
Approximately how much is this per gallon?

7 Carol weighs 126 pounds. Jane weighs 132 pounds.
Approximately how many kilograms heavier is Jane than Carol?

8 Karen cycles 20 miles around a forest trail.
Approximately how many kilometres is this?

9 Rebecca drives 42 km to work.
How many miles is this? Give your answer to the nearest mile.

E

10 The speed limit on motorways in the UK is 70 mph. The speed limit on motorways in France is 120 km/h. Jack says that you can drive faster on French motorways. Jill says that UK motorways have a faster speed limit.
Who is correct? Show your working.

11 Bill enters a 10 km race. He says it is the same distance as the 6 mile training runs he has been doing. Is Bill correct? Show your working.

12 Iqbal says that 40 miles is the same as 25 km.

   a What mistake has Iqbal made?

   b What is 40 miles in kilometres?

   c What is 25 kilometres in miles?

13 Janet, Liz and Peter are doing a sponsored walk.

The walk is 24 km.

Janet will raise £400 if she completes the walk.

Peter will get £30 for **each mile** he walks.

They are hoping to raise £1000 for charity. They all complete the walk.

How much does Liz need to raise for each mile she walks?

14 Fran buys milk at her corner shop. She pays 96p for a container which holds 2 pints. The mini-market next door sells milk in containers which hold 1 litre for 85p. There are 8 pints in 1 gallon.

Which shop sells milk at the cheaper price? You must show your working.

Learn... **12.4 Compound measures**

Compound measures combine two different units. For example,

$$\text{speed} = \frac{\text{distance}}{\text{time}} \qquad \text{fuel consumption} = \frac{\text{number of miles travelled}}{\text{number of gallons used}}$$

Be careful with the units. They give you clues about what to divide by.

Speed is measured in kilometres per hour (km/h).

This tells you the formula:

$$\text{speed} = \frac{\text{distance (km)}}{\text{time (hours)}}$$

Fuel consumption is measured in miles per gallon or kilometres per litre.

> **Bump up your grade**
>
> Make sure you have the correct units for the problem. To answer questions at Grade C you may need to convert some units before you use them.
>
> For km/h you need the distance to be in km and the time to be in hours.

Example: Jamie is a runner in a club. He runs 200 metres in 32 seconds. Find his average speed in:

  a metres per second

  b kilometres per hour.

Solution: **a** Using the formula above:

$$\text{average speed} = \frac{\text{distance}}{\text{time}} = \frac{200}{32} \text{ m/s} = 6.25 \text{ m/s}$$

> **Hint**
>
> metres per second
> = metres/second
> = metres ÷ seconds

  b 6.25 m/s means 6.25 metres every second.

  In 60 seconds he runs $6.25 \times 60 = 375$ metres

  In 60 minutes he runs $375 \times 60 = 22\,500$ metres

  $22\,500$ metres $= 22\,500 \div 1000 = 22.5$ km

  Jamie runs at a speed of 22.5 km/h.

Example: Work out the average speed in mph of:

  a a train that takes $1\frac{1}{2}$ hours to travel 102 miles

  b a plane that travels 1450 miles in 2 hours 45 minutes.

> **Hint**
>
> mph = miles per hour
> = miles/hour
> = miles ÷ hours

Solution: **a** Using the formula above:

$$\text{average speed} = \frac{\text{distance}}{\text{time}} = \frac{102}{1.5} = 68 \text{ mph}$$

  b Using the formula, average speed $= \dfrac{\text{distance}}{\text{time}}$

  $$\text{Average speed} = \frac{1450}{2.75} = 527 \text{ mph}$$

> **AQA Examiner's tip**
>
> Change the time to hours, using decimals, when the time is in hours and minutes and the answer is in mph.

Practise... **12.4 Compound measures** **G F E D C**

1 Work out the average speed for each of the following.
State the units of your answers.

  a A car travels 200 metres in 8 seconds.

  b A man takes 28 seconds to run 200 metres.

  c A train takes 2 hours to travel 230 miles.

2 Express each of these times as decimal fractions of an hour.

   a   30 minutes

   b   15 minutes

   c   4 hours 45 minutes

3 Write each of these times as hours and minutes.

   a   2.5 hours

   b   3.25 hours

   c   1.75 hours

4 Find the speed in mph of:

   a   a car that travels 85 miles in 1 hour 15 minutes

   b   a lorry that travels 75 miles in 1 hour 30 minutes.

> **AQA** *Examiner's tip*
>
> Remember to divide by time in hours when you find a speed in miles per hour (mph) or kilometres per hour.

5 A snail crawls at 5 cm per minute.

   a   How far does it crawl in 1 hour?

   b   How long does it take to crawl 1 metre?

6 Work out the time taken for each of these journeys.
Give your answer in hours and minutes.

   a   A car travels 40 km at 50 km per hour.

   b   A bus travels 20 km at 30 km per hour.

   c   A cyclist travels 45 km at 25 km per hour.

7 Work out the distance travelled for each of these journeys.

   a   A person walks at 4 km per hour for 75 minutes.

   b   A train travels at 110 km per hour for 90 minutes.

   c   A lorry travels for 45 minutes at 50 km per hour.

8 Jan drives 255 miles in her car and uses 6 gallons of fuel.

   a   What is her fuel consumption in miles per gallon?

   b   How many gallons of fuel would Jan use for a similar journey of 400 miles?

   c   The fuel tank in Jan's car holds 15 gallons of fuel when it is full.

   Is it possible for Jan to travel 600 miles on one full tank of fuel?

9 John runs at 9 km per hour for 40 minutes. He then walks 2.5 km in 30 minutes.
What is John's average speed in km per hour over the whole journey?

10 Sam drives 265 miles from Kendal to Bristol.

His average fuel consumption for this journey is 50 miles per gallon (mpg).

His fuel tank holds 15 gallons of fuel when full.

He then drives 221 miles from Bristol to Norwich with an average fuel consumption of 48 mpg.

Lastly, he drives 278 miles from Norwich back to Kendal, with an average consumption of 47 mpg.

Is it possible for Sam to complete this journey on one tank of petrol?

Learn...   **12.5 Accuracy in measurements**

When you measure in centimetres and millimetres, any length between 6.5 cm and 7.5 cm will round to 7 cm.

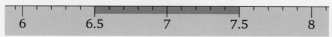

> **AQA** *Examiner's tip*
>
> 6.5 cm to 7.5 cm means any value in the range 6.5 up to but not including 7.5

On this scale any measurement in the shaded area rounds to 7 cm.

The shaded area is from 6.5 cm up to 7.5 cm.

Notice: 'to the nearest cm' means the actual distance could be any value from half a centimetre below to half a centimetre above.

Example: Zac measures the length of a shelf. He says it is 43 cm to the nearest centimetre. What are the smallest and the largest possible lengths of the shelf?

Solution: Zac's measurement rounds to 43 cm, so the actual length must be in the shaded area on the scale.

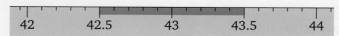

The shaded area is from 42.5 cm up to 43.5 cm .

The smallest length is 42.5 cm.

The largest length is 43.5 cm.

Practise...   **12.5 Accuracy in measurements**   k!    G F E D C

C

1   Round each of the following to the nearest whole number.

   a   12.9 cm   **c**   49.9 miles   **e**   43.56 kg

   b   81.2 m   **d**   6.5 g   **f**   47.4999 litres

2   Each of these lengths rounds to 15 cm, to the nearest centimetre.

     14.6 cm, 15.4 cm, 14.91 cm, 15.49 cm

   a   Write down some other lengths that also round to 15 cm to the nearest cm.

   b   What is the minimum value of a measurement which has been rounded to 15 cm?

   c   What is the maximum value of a measurement which has been rounded to 15 cm?

3   George measured the length of a post for a washing line in his garden. He found it was 3 metres to the nearest metre.

   What are the smallest and largest possible lengths of the post?

4   Faye measured some of the crayons in her pencil case. She found they were all 8 cm to the nearest cm.

   Does this mean they were all the same length? Give a reason for your answer.

5   Mike measured the distance from his home in Windermere to his school in Kendal. He found it was 18 km to the nearest kilometre.

   a   What is the smallest distance that Mike's measurement could be?

   b   What is the largest distance that Mike's measurement could be?

C

6 The weight of a letter is 43 g to the nearest gram.

What are the smallest and largest values for the weight of the letter?

7 A bridge has a sign stating the maximum weight allowed on it is 2 tonnes.

A van driver knows his van weighs 2 tonnes to the nearest tonne.

Can the van driver be sure that it is safe for him to drive over the bridge? Give a reason for your answer.

8 The contents of a packet of crisps weigh 33 g to the nearest gram.

Which of the following could be the weight of the crisps?

a	33.2 g	c	33.49 g	e	32.5 g
b	33.6 g	d	33.94 g	f	32.29 g

⚠ 9 Rachel weighs the books in her school bag.
All her textbooks together weigh 3 kg to the nearest kilogram.
All her exercise books together weigh 1 kg to the nearest kilogram.
There are no other books in her bag.
What is the maximum weight of the books in her bag?

⚠ 10 Mark buys some food from the grocer's shop. He buys 1 kg of carrots, 5 kg of potatoes, both weighed to the nearest kilogram.
What is the minimum possible weight of his shopping?

⚙ 11 George, Mildred and Henrietta all get into a lift.
George weighs 95 kg, Mildred weighs 83 kg and Henrietta weighs 71 kg.
The lift states: Maximum weight 250 kg.
Is it safe for George, Mildred and Henrietta to use the lift?
Does it make a difference if the weights have been rounded? Give a reason for your answer.

12 Assess ⓚ!

G

1 Measure the following lines. Record your measurement to the nearest centimetre.

2 What are the most appropriate units for measuring each of the following?

a The mass of a car

b The height of a house

c Your mass

d The mass of a bag of sweets

e The capacity of a teaspoon

G

3 What number does each arrow point to?

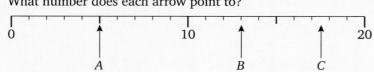

4 A bottle contains 2 litres of water.
How many millilitres is this?

5 Sam trains at the athletic track. He completes 5 laps of a 400 metre track.
How far does Sam run? Give your answer in kilometres.

F

6 Jack buys a 4 kg bag of potatoes from his local shop.
What is the mass of the potatoes in pounds?

7 James buys a 4 pint bottle of milk.
How many litres of milk are in the bottle?

E

8 Frank is driving his car in France. The speed limit is 110 km per hour.
What is this in miles per hour?

D

9 A van travels 230 miles in 7 hours.
What is the average speed of the van?

C

10 A car takes 45 minutes to travel 17 miles.
What is the average speed of the car?

11 The temperature in a classroom is 18°C to the nearest degree.
What are the highest and lowest temperatures that would round to 18°C?

AQA Examination-style questions

1 Dipak travels a distance of 30 miles.
Wendy travels a distance of 40 kilometres.

Who travels further?
You **must** show your working.

(3 marks)

AQA 2008

Trial and improvement

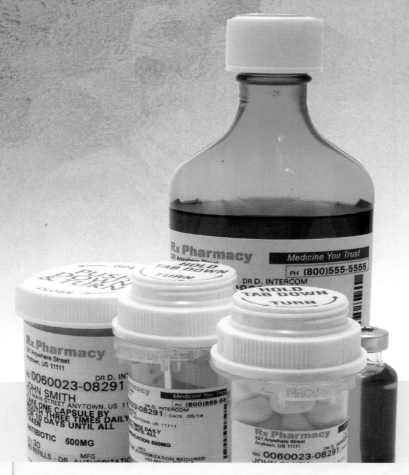

Examiners would normally expect students who get these grades to be able to:

C

solve equations such as $x^3 + x = 12$ using systematic trial and improvement methods.

Did you know?

Trial and improvement

Trial and improvement (also know as trial and error) has many uses in the real world. It is often used by engineers when they develop complex equipment. For example, engineers will trial different fuel flow rates when determining the maximum thrust from a jet engine.

Also doctors may use trial and improvement to test different combinations of drugs for diabetes, epilepsy and high blood pressure. For the future, scientists are developing supercomputers which will act as 'virtual humans'. This will allow doctors to match different combinations of drugs to different patients.

You should already know:

✔ how to substitute into algebraic expressions

✔ how to rearrange formulae

✔ how to use the bracket and power buttons on your calculator.

 **Learn...** **13.1 Trial and improvement**

Trial and improvement is a method for solving problems using estimations which get closer and closer to the actual answer. Trial and improvement is used where there is no exact answer so you will be asked to give a rounded answer. On the examination paper you will be told when to use a trial and improvement method.

If you are told that $x^3 + x = 50$ then you can work out that the answer lies between 3 and 4 because:

$$3^3 + 3 = 30 \quad \text{which is too small}$$

and $\quad 4^3 + 4 = 68 \quad$ which is too large.

You know that the answer lies between 3 and 4 so you might try 3.5

$$3.5^3 + 3.5 = 46.375 \quad \text{This is too small.}$$

As 3.5 is too small and 4 is too large, you know that the answer lies between 3.5 and 4 so you might try 3.7 or 3.8.

You can keep going with this method to get an answer which is more and more accurate.

The question will tell you how accurate your answer should be.

> **AQA** *Examiner's tip*
>
> It is a good idea to lay out your working carefully. A table can be helpful.

Example: Use trial and improvement to solve $x^3 - x = 40$

Give your answer to one **decimal place**.

Solution: You can try out some different values to get you started.

Trial value of x	$x^3 - x$	Comment
1	0	Too small
2	6	Too small
3	24	Too small
4	60	Too large

40

> **AQA** *Examiner's tip*
>
> On some examination questions you will be told where the answer lies. For example, you may be told that there is a solution between 2 and 3.

The answer 40 lies between 24 and 60.

This tells you that x lies between 3 and 4.

You might try 3.5

3.5	39.375	Too small
3.6	43.056	Too large

Again, you can see that 40 lies between 39.375 and 43.056

This tells you that x lies between 3.5 and 3.6

You should try 3.55

The answer to one decimal place is either 3.5 or 3.6

Work out the value for 3.55 to see if it is larger than 40.

If it is too large, then 3.55 is too large and the answer to one decimal place is 3.5

If it is too small, then 3.55 is too small and the answer to one decimal place is 3.6

3.55	41.188875	Too large

You know that x lies between 3.5 and 3.55

But any answer between 3.5 and 3.55 is the same as 3.5 to one decimal place.

The required answer is 3.5 to one decimal place.

Practise... 13.1 Trial and improvement

C

1 Mavis the farmer knows the area of her square fields but not the length.
Use trial and improvement to find the length of the fields.
Give your answer to two decimal places.

a
Area
$120\,\text{m}^2$

b
Area
$600\,\text{m}^2$

c
Area
$320\,\text{m}^2$

2 Find, using trial and improvement, a solution to the following equations.
Give your answers correct to one decimal place. Remember to show all your working.

a $x^3 + x = 10$ **c** $x^3 - 5x = 400$

b $x^3 + x = 520$ **d** $x - x^3 = 336$

3 Use trial and improvement to solve the equation $x^3 + x = 75$
Give your answer to one decimal place.
The table has been started for you.

Trial value of x	$x^3 + x$	Comment
2	10	Too low
3	30	Too low
4	68	Too low
5	130	Too high

So now we know that the value lies between 4 and 5.

4.5	95.625	Too high
?		
?		

4 Use trial and improvement to find solutions to the following equations.
Give your answer to two decimal places.

a $a^3 - 10a = 50$ if the solution lies between 4 and 5

b $x^3 - x = 100$ if the solution lies between 4 and 5

c $5x - x^3 = 10$ if the solution lies between -2 and -3

5 Use trial and improvement to find solutions to the following equations.
Give your answer to two decimal places.

a $t^3 - 5t = 10$ if t lies between 2 and 3

b $x^3 - 5x = 60$ if x lies between 4 and 5

c $x(x^2 + 1) = 60$ if x lies between 3 and 4

d $p^3 + 6p = -50$ if p lies between -3 and -4

6 Use trial and improvement to find a negative solution to the equation $y^3 + 60 = 0$

7 The equation $x^3 - 4x^2 = -5$ has two solutions of x between 0 and 5.
Use trial and improvement to find these solutions.
Give your answer to three decimal places.

⚠ **8** Use trial and improvement to find the value of $x^2 - \dfrac{1}{x} = 5$ where x lies between 2 and 3.

Give your answer to two decimal places.

? **9** The difference between the square of a number and the cube of a number is 100.

Find the number to one decimal place.

? **10** This solid consists of a central square and four equal arms.

The volume of the solid is $100\,\text{cm}^3$.

Find the value of x correct to two decimal places.

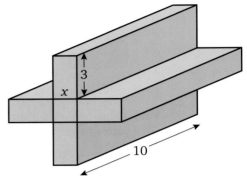

13 Assess (k!)

C **1** Use trial and improvement to solve the equation $x^3 - 4x = 100$
Give your answer to one decimal place.
The table has been started for you.

Trial value of x	$x^3 - 4x$	Comment
3	15	Too small
4	48	Too small
5	105	Too large

2 The equation $x^3 + 8x^2 = 20$ has two negative solutions between 0 and -8.

Use trial and improvement to find these solutions.

Give your answer to two decimal places.

3 A cuboid measures $x \times x \times (x + 2)$.

The volume of the cuboid is $50\,\text{cm}^3$.

Use trial and improvement to find x.

Give your answer to three decimal places.

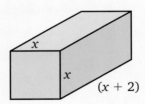

4 Use trial and improvement to find solutions to the following equations.

Give your answer to two decimal places.

a $x^3 + 5x = 50$ if the solution lies between 3 and 4

b $y^3 + 3y = 10$ if the solution lies between 1 and 2

c $x^3 - x = 100$ if the solution lies between 4 and 5

d $3x - x^3 = 25$ if the solution lies between -3 and -4

5 Use trial and improvement to find solutions to the following equations.

Give your answer to two decimal places.

a $x^3 + x = 60$ if x lies between 3 and 4

b $x^3 - 12x = 0$ if x lies between 2 and 6

6   Use trial and improvement to find the value of $8x - x^3 = 3$ where x is negative.

Give your answer to two decimal places.

7   Use trial and improvement to find the value of $t^3 + t^2 = 10$ where $1 \leqslant t \leqslant 2$

Give your answer to two decimal places.

AQA Examination-style questions

1   Kerry is using trial and improvement to find a solution to the equation $8x - x^3 = 5$
Her first two trials are shown in the table.

x	$8x - x^3$	Comment
2	8	too high
3	−3	too low

Copy and continue the table to find a solution to the equation.
Give your answer to one decimal place.

(3 marks)

AQA 2007

2   The sketch shows the graph of $y = x^3 - 3x - 8$
The graph passes through the points $(2, -6)$ and $(3, 10)$.

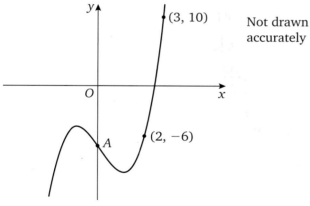

Not drawn accurately

a   The graph crosses the y-axis at the point A.
Write down the coordinates of point A.

(1 mark)

b   Use trial and improvement to find the solution of:
$$x^3 - 3x - 8 = 0$$
Give your answer to one decimal place.

(4 marks)

AQA 2008

14 Enlargements

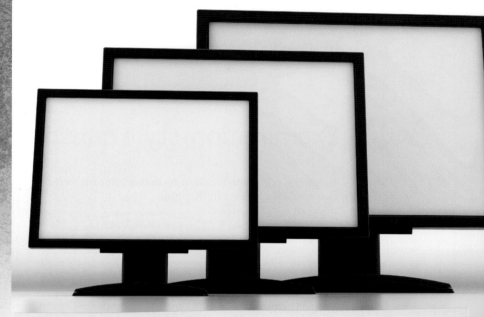

Examiners would normally expect students who get these grades to be able to:

F

state the scale factor of an enlargement

E

enlarge a shape by a positive scale factor

find the measurements of the dimensions of an enlarged shape

D

enlarge a shape by a positive scale factor from a given centre

C

find the ratio of corresponding lengths in similar shapes and identify this as the scale factor of enlargement

use ratios in similar shapes to find missing lengths.

Key terms

enlargement
transformation
similar
scale factor
centre of enlargement
vertex, vertices
ratio

Did you know?

Screen sizes

The screens for monitors or televisions come in many different sizes. The two larger screens in the photograph above look as though they are enlargements of the smaller screen. Enlargements have to satisfy certain conditions which you will learn about in this chapter.

Television screens come in two different shapes: 4 : 3 and 16 : 9

- If you measure an older style TV, the 4 : 3, you will find that for every 4 units of width there are 3 units of height.
- If you measure a widescreen TV screen, which is 16 : 9, for every 16 units of width there are 9 units of height.

An enlargement of 4 : 3 with a scale factor of 4 would be 4 × 4 : 4 × 3 which equals 16 : 12. Therefore the ratio 16 : 9 is not an enlargement of 4 : 3. This is why people filmed in 4 : 3 format look fatter when shown in widescreen.

Strangely, televisions are measured across the diagonal and still in inches!

You should already know:

- ✓ how to plot coordinates in all four quadrants
- ✓ about units of length and how to use them
- ✓ about ratio and how to simplify a ratio
- ✓ how to use the vocabulary of transformations: mapping, object and image
- ✓ how to recognise and use corresponding angles
- ✓ how find the area of simple shapes including a rectangle and a triangle.

14.1 Introduction to enlargement and scale factor

Enlargements are a type of **transformation**.

They are the only transformations at GCSE that change the size of a shape.

All the other transformations (reflections, rotations and translations) keep the image the same size as the original shape. The shapes are **congruent**.

An enlargement changes the size of an image but not the shape.

All the lengths will be changed but all the angles will stay the same. The shapes are **similar**.

For example, if you take a photograph to be enlarged, the new photograph will be bigger but the picture does not change in any other way.

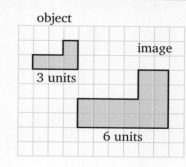

Example: These two diagrams are the same shape.
The shapes are similar.
One is an enlargement of the other.

What is the scale factor of the enlargement?

Solution: The scale factor tells us how many times bigger the shape has been made. To find it you need to take two corresponding sides, one on the original (object) and one on the new one (image).

In this diagram the corresponding lengths are 3 units and 6 units. The length on the image is 2 times that on the object.

So, the scale factor = 2

Every length on the image is twice the size of the corresponding length on the object.

Example: Copy this diagram onto squared paper and then enlarge the shape by scale factor 3.

Solution: Before you start, look at the lengths of the lines already given on the original shape.

As they are to be enlarged by a scale factor of 3, each of these lengths will be three times longer in the image.

Starting at the top left-hand side of the shape, the top is 3 units in length.

The top of the enlarged shape will be 3 × 3 = 9 units long.

Move around the shape and enlarge each side in turn.

The side was 1 unit long originally so on the enlarged shape it will be 1 × 3 = 3 units long.

Continue in this way until you get back to the starting point.

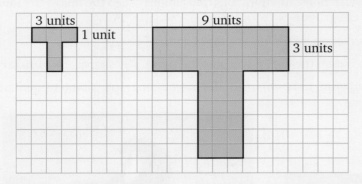

14.1 Introduction to enlargement and scale factor

Practise...

G F E D C

F

1 In these diagrams, *A* has been enlarged to give *B*.

What is the scale factor of each enlargement?

a 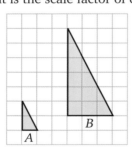   **b** 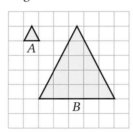   **c**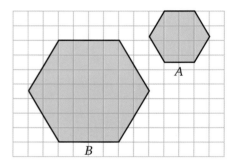

E

2 A trapezium *ABCD* is enlarged to form the trapezium *EFGH*.

The scale factor of the enlargement is 4.

AB = 2.5 cm   *DC* = 2 cm   *AD* = 1.5 cm   *BC* = 1.5 cm

Work out the lengths of the sides:

a *EF*   **b** *HG*   **c** *EH*   **d** *FG*

Not drawn accurately

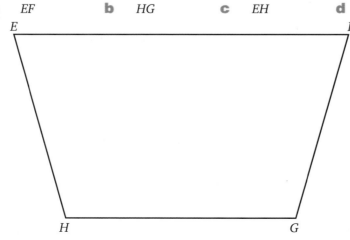

Not drawn accurately

Hint

EF means the line joining *E* and *F*.

3 Enlarge each of these shapes with a scale factor 2.

a

b

c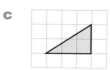

4 a Enlarge the shape with a scale factor of 3.

b A similar triangle to *ABC* was enlarged, using the same scale factor, to give an image triangle *A'B'C'*.

If *A'B'* was 6.708 metres (to 3 d.p.), what was the length of *AB* (to 3 d.p.)?

⚠ 5 Triangle *A* is enlarged with scale factor 3 to give triangle *B*.

a One side of triangle *B* has length 7.5 cm.

What is the length of the corresponding side of triangle *A*?

b One angle of triangle *B* is 45°.

What is the size of the corresponding angle in triangle *A*?

⚠ 6 The diagram shows the plan for a garden. It consists of a lawn with three square flower beds.

The smallest flowerbed, *X*, is 1.4 metres by 1.4 metres.

a *Y* is an enlargement of *X* with a scale factor of 2.
Write down the dimensions of *Y*.

b *Z* is an enlargement of *Y* also with a scale factor of 2.
Write down the dimensions of *Z*.

c Write down the scale factor of the enlargement that takes *X* to *Z*.

d Find the area of each flowerbed.
Give your answers to two decimal places.

e Find the area of the lawn.
Give your answer to two decimal places.

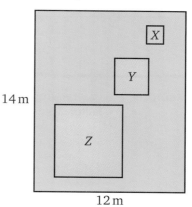

14 m

12 m

⚙ 7 a Julie has a photograph of her cat. She wants to have the photograph enlarged to put on the cover of her portfolio.

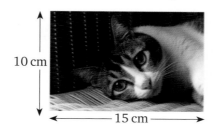

10 cm

15 cm

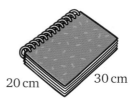

20 cm   30 cm

She wants the picture to fill the cover of the portfolio.
What is the scale factor of the enlargement she needs?

b Julie also wants to enlarge the photograph to fit a frame to go on her wall.
The frame is 90 cm wide.

i Find the scale factor to be used to enlarge the original photograph to fit the frame.

ii Find the height of the frame.

iii If the photograph on the folder was enlarged to fit the frame, find this scale factor.

Learn... 14.2 Centres of enlargement

When diagrams are drawn on sets of axes, extra information is needed to perform an enlargement.

The **centre of enlargement** is given as a pair of coordinates.

This tells you where to put the enlargement on the axes.

The centre of enlargement can be anywhere including inside, outside or on the edge of the object.

In all diagrams shown here, the same scale factor has been used but the centre of enlargement, marked with a cross, is different.

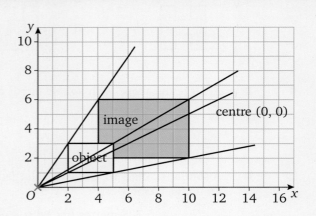

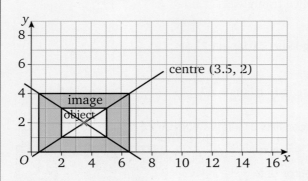

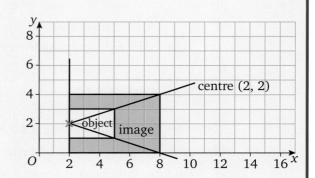

Example: Enlarge the rectangle *A* by scale factor 2, centre of enlargement (1, 1).

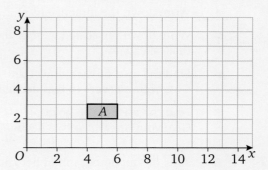

Solution: To enlarge a shape:

1. Plot the centre of enlargement on the grid with a cross.

2. Choose a **vertex** (corner) of the shape. Join the centre of enlargement to this vertex and extend the line past the vertex.

3. Measure the distance from the centre of enlargement to the vertex and multiply this by the scale factor. In this case, the scale factor is 2.

4. This is the new distance from the centre of enlargement to the corresponding vertex on the new rectangle. Measure this distance along the line you have drawn and mark the new point.

5. Repeat this for all other **vertices**.

AQA Examiner's tip

Always use a sharp pencil and a ruler. Make sure that the lines are drawn exactly through the intersection of the lines of the grid.

AQA Examiner's tip

There are usually two vertices that are easier to draw because the construction lines do not cross over the shape itself.
Do these first!

When the enlargement is finished, the distances from the vertices to the point of enlargement will be twice as long.

Every length on the new rectangle will be twice as long as it was before.

The original rectangle was 2 by 1 units. The enlarged rectangle is now 4 by 2 units.

The rectangles are similar.

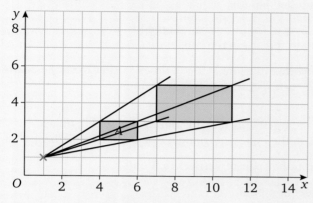

An alternative method would be to count the horizontal and vertical distance from the centre of enlargement to a vertex.

The new vertex would be at a point twice the horizontal distance and twice the vertical distance from the centre of enlargement.

Hint

You can use the scale factor to check the dimensions of the enlarged rectangle. You can use this fact to help you draw an accurate diagram and to check your enlargement.

AQA Examiner's tip

You can use either of these methods to find the new position of a vertex. You can even use a combination of both!

Example: The triangle A has been enlarged to triangle B.

Find:

a   the scale factor of enlargement

b   the centre of enlargement.

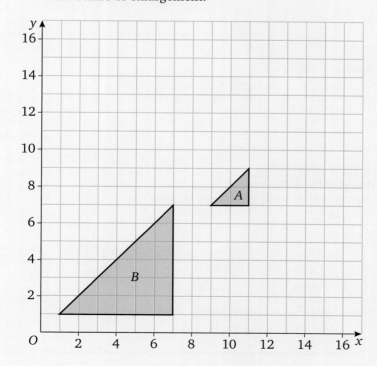

Solution: **a**   Measure a side on the enlarged shape *B* and a corresponding side on the original shape *A*.

For example the 'bottom line' of *B* measures 6 units and the 'bottom line' of *A* measures 2 units.

The scale factor $= \dfrac{\text{enlarged length}}{\text{original length}} = \dfrac{6}{2} = 3$

The scale factor is 3. Every length on the enlarged triangle is 3 times the corresponding length on the original triangle.

The distance of each vertex in triangle *B* is 3 times further from the centre of enlargement than the corresponding vertex in triangle *A*.

b   To find the centre of enlargement, start by joining a vertex in object *A* with the corresponding vertex in the enlarged triangle *B*. Extend the line back past *A*.

Now do the same for anther pair of vertices in *A* and *B*.

The two lines will meet at a point.

This point is the centre of enlargement.

For this enlargement, the centre of enlargement is (13, 10).

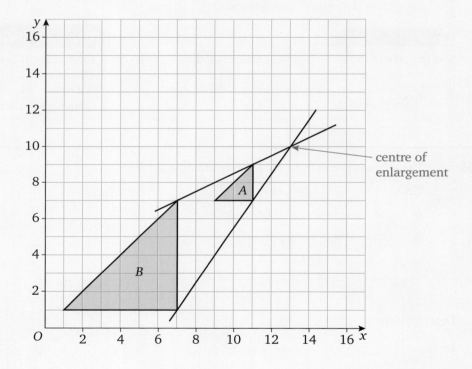

AQA **Examiner's tip**

Once you have identified the centre of enlargement, choose a vertex.

Then check that:

the distance of the vertex on the enlarged image from the centre of enlargement = (scale factor) × distance of the corresponding vertex in the original object from the centre of enlargement.

Practise... 14.2 Centres of enlargement

1 Copy each shape onto squared paper. (You will need a 12 by 12 grid for each part of the question.)

Enlarge each shape with scale factor 2 and centre of enlargement X.

Use the grid position marked with a cross as the centre of enlargement.

a

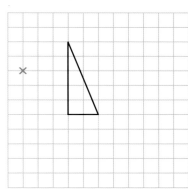

b

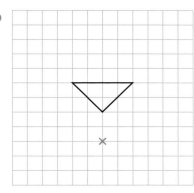

c

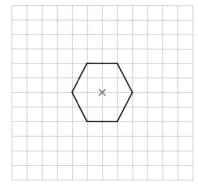

2 Each part of this question shows the same shape but the centre of enlargement is in a different place.

Draw a separate diagram showing the enlargement for each part of the question.

You will need a 12 by 12 grid for each one. The scale factor of enlargement is 3.

a

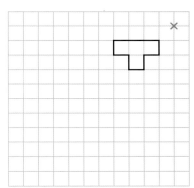

b

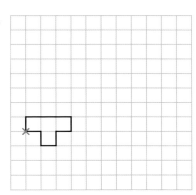

c

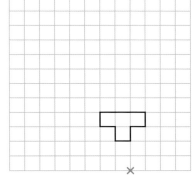

3 Describe each of the enlargements.

In each diagram, A has been enlarged to give B.

> **Hint**
>
> To describe an enlargement fully you need to give the scale factor and centre of enlargement.

a

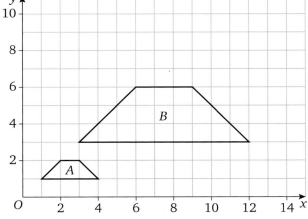

b

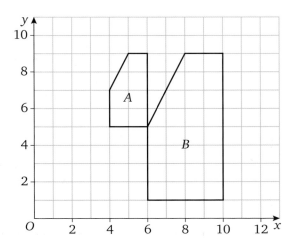

D

c

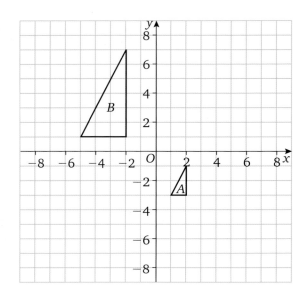

4 Describe fully the enlargement of triangle *ABC* to triangle *DEF*.

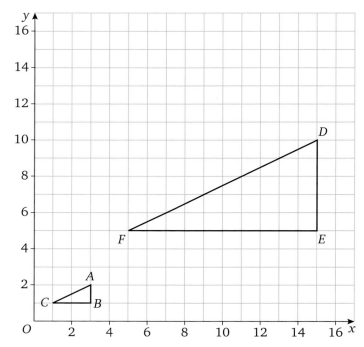

5 a i Draw a pair of axes with values from −9 to 7.

ii Plot and label triangle *B* with vertices at (−4, 1), (−4, −4) and (−6, −1).

b Draw the image of triangle *B* after an enlargement of scale factor 3, with centre of enlargement (−9, −2).
Label the image *C*.

c What are the coordinates of the vertices of *C*?

6 a i Draw a pair of axes with *x*-values from 0 to 16 and *y*-values from 0 to 8.

ii Plot and label rectangle *E* with vertices at (6, 2), (6, 4), (10, 4) and (10, 2).

b Draw the image of rectangle *E* after an enlargement of scale factor 1.5, with centre of enlargement (0, 0).
Label the image *F*.

c What are the coordinates of the vertices of *F*?

Bump up your grade

For a Grade C you need to know how to work in all four quadrants.

7   A jewellery designer was asked to make a pendant which perfectly matched a pair of earrings. In order for them to look similar, he decided that the pendant should be an enlargement of the earrings. He worked out that a scale factor of 3 would be most suitable.

Here is the plan of one of the earrings.
Each square represents a square 0.5 cm by 0.5 cm.

Copy this diagram onto squared paper and draw the plan for the pendant.

The centre of enlargement is marked by a cross on the diagram.

Learn... 14.3 Enlargements, similar shapes and ratio

Triangle *ABC* is an enlargement of triangle *DEF* with a scale factor of 2. This means that the triangles are similar.

Every side on triangle *ABC* is twice the length of the corresponding side on triangle *DEF*. The corresponding angles are equal.

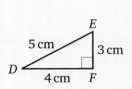

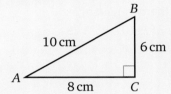

Not drawn accurately

The corresponding sides are all in the same **ratio** 2 : 1

$AC : DF = 8 : 4$ Divide both sides by 4.
$\quad\quad\quad = 2 : 1$

$BC : EF = 6 : 3$ Divide both sides by 3.
$\quad\quad\quad = 2 : 1$

and

$AB : DE = 10 : 5$ Divide both sides by 5.
$\quad\quad\quad = 2 : 1$

Ratios can be simplified or cancelled down in the same way as fractions.

This is done by dividing both sides by the same number until you cannot do it any more. It is then in its simplest form.

Example: Rectangle *EFGH* is an enlargement of rectangle *ABCD*.

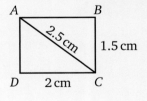

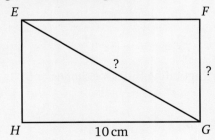

Not drawn accurately

  a   Write down the ratio of *DC : HG* in its simplest form.

  b   What is the scale factor of the enlargement?

  c   Find the length of *FG*.

  d   Find the length of the diagonal *EG*.

  e   Work out the perimeter of each rectangle.

  f   What is the ratio of the perimeter of *ABCD* to the perimeter of *EFGH*?

Solution:

a $DC = 2$ cm $HG = 10$ cm

     $DC : HG = 2 : 10$

           $= 1 : 5$ Divide both sides by 2

The scale factor is the value of n when the ratio of the corresponding lengths is written in the form $n : 1$ or $1 : n$.

b   The scale factor is taken from the ratio once it is in the form $1 : n$ or $n : 1$

     Scale factor of the enlargement $= 5$

c $BC = 1.5$ cm

     $FG = 1.5 \times 5 = 7.5$ cm

d $AC = 2.5$ cm

     $EG = 2.5 \times 5 = 12.5$ cm

e   Perimeter of $ABCD = 2 + 1.5 + 2 + 1.5 = 7$ cm

     Perimeter of $EFGH = 10 + 7.5 + 10 + 7.5 = 35$ cm

f   Comparing the perimeters of the two rectangles:

     perimeter of $ABCD$: perimeter of $EFGH = 7 : 35$

                                     $= 1 : 5$ Divide both sides by 7.

This is the same ratio as before.

Perimeter is also a length.

The ratios of all corresponding lengths in the diagram should be the same because the shapes are similar.

> **AQA Examiner's tip**
>
> Whenever you are asked to find missing lengths, always check afterwards that the ratio of the original length to the enlarged length is correct.

Example: Are these shapes similar? Is shape A an enlargement of shape B?

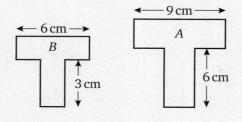

Not drawn accurately

Solution: Ratio of 'tops' of shapes $= 9 : 6$ Divide both sides by 3

                             $= 3 : 2$ Divide both sides by 2

                             $= 1.5 : 1$

> **Hint**
>
> To read off a scale factor from a ratio, the ratio must be in the form $n : 1$ or $1 : n$

Scale factor $= 1.5$

Ratio of corresponding 'sides' of shapes $= 6 : 3$ Divide both sides by 3

                                        $= 2 : 1$

Scale factor $= 2$

The scale factors are different, so these shapes are not similar. A is **not** an enlargement of B.

> **Bump up your grade**
>
> For a Grade C you should be able to express enlargements using scale factors and also by ratios. You should be able to understand the connection between them and use them to find missing lengths.

14.3 Enlargements, similar shapes and ratio

D

1 Triangles A and B are equilateral.

a What is the ratio of the lengths?
Give your answer in its simplest form).

b Find the perimeter of each equilateral triangle.

c What is the ratio of the perimeters?
Give your answer in its simplest form.

d What is the scale factor of the enlargement?

Not drawn accurately

2 The large triangle is an enlargement of the small triangle.

a What is the ratio of the bases of the triangles?
Give your answer in its simplest form.

b What is the scale factor of the enlargement?

c Find the missing height of the enlarged triangle.

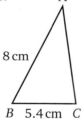

Not drawn accurately

3 Triangle DEF is an enlargement of triangle ABC.

a Work out the ratio of the length of AB to the length of DE. Give your answer in the form 1 : n.

b Find the length of the side EF.

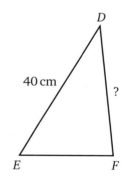

Not drawn accurately

C

4 In the following diagrams, decide if the larger shape is an enlargement of the smaller one.

In both parts, all corresponding angles are equal.

a

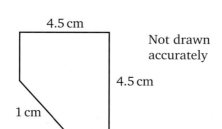

Not drawn accurately

b

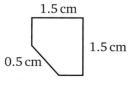

Not drawn accurately

C

5 These triangles are enlargements of each other. Work out the missing lengths.

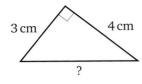

3 cm   4 cm

?

Not drawn accurately

36 cm   ?

60 cm

6 These triangles are enlargements of each other.

3 cm   2.4 cm   3 cm

?

Not drawn accurately

24 cm   ?   ?

28.8 cm

a   Find the ratio of the corresponding sides.

b   Find any missing lengths.

7 Here is a diagram of some stepladders.

There are two similar triangles in this diagram.

a   Draw them out as two separate diagrams.

b   Transfer the measurements onto these diagrams.

c   What is the length of the rope?

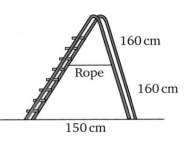

160 cm   Not drawn accurately

Rope

160 cm

150 cm

8

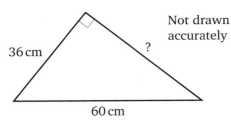

Richard wants to know if an enlargement of the photograph will fit the frame. The photo has dimensions 15 cm by 10 cm.
The frame measures 42 cm by 28 cm.

a   Are the shapes similar? If so, what is the scale factor?

b   What would be the size of the frame needed if the photograph had been enlarged by scale factor 4?

9 Russian wooden dolls are made so that one doll will fit completely inside the next one.

They are similar.

By measuring the heights of the dolls in the picture, see if you can find an approximate scale factor relating the first doll to the second, the second to the third and so on.

Are all the scale factors the same?

14   Assess (k!)

1   State the scale factor of the following enlargement.

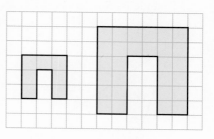

2   Copy this shape and draw an enlargement of the shape, scale factor 3.

On the enlarged shape, what are the lengths of the two parallel sides?

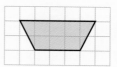

3   Copy this shape and enlarge it with scale factor 2 about the centre of enlargement shown.

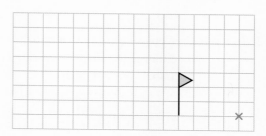

4   **a**   **i**   Draw a pair of axes with x- and y-values from 0 to 16.

   ii   Plot and label triangle A with vertices (6, 7), (6, 11) and (10, 11).

   b   Draw the image of triangle A after an enlargement of scale factor 3, with centre of enlargement (7, 10).
   Label this triangle B.

   c   What are the coordinates of the vertices of B?

5   A tennis court is an enlargement of a table tennis table.

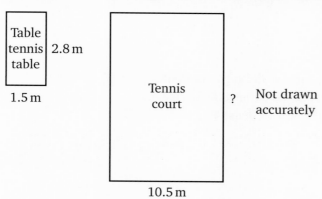

The tennis court is similar to the table-tennis table.
One is an enlargement of the other.

   a   Work out the scale factor of the enlargement.

   b   Work out the length of the tennis court.

   c   Work out the area of the table-tennis table and the area of the tennis court.

   d   Work out the ratio of the area of the table tennis table to the area of the tennis court.
   Give your answer in the form $1 : n$.

AQA Examination-style questions ⓚ⁴

1 Triangle *PQR* is an enlargement of triangle *ABC* with scale factor 5.

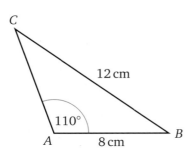

Calculate the length of *RQ*. *(2 marks)*

AQA 2007

2 **a** **i** Copy shape *A* onto a grid.
      ii Enlarge the shape by a scale factor of 2.
      iii Label your new shape *B*. *(2 marks)*

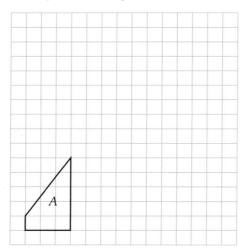

  b Which of these statements are true?
   i Shape *B* is congruent to shape *A*.
   ii The angles of shape *B* are the same as the angles of shape *A*.
   iii The perimeter of shape *B* is twice the perimeter of shape *A*.
   iv The area of shape *B* is twice the area of shape *A*. *(4 marks)*

AQA 2006

15 Construction

Objectives

Examiners would normally expect students who get these grades to be able to:

G

select congruent shapes

measure a line accurately to the nearest millimetre

F

use simple scale drawings

measure and draw an angle to the nearest degree

E

understand congruence and similarity

use scales, such as a scale on a map

draw scale drawings

draw a triangle given three sides, or two sides and the included angle, or two angles and a side

D

draw a quadrilateral such as a kite, parallelogram or rhombus with given measurements

understand that giving the lengths of two sides and a non-included angle may not produce a unique triangle

C

construct perpendicular bisectors and angle bisectors.

Did you know?

This type of mathematics is part of geometry and is very old

The ancient Greek mathematician Euclid is the inventor of geometry. He did this over 2000 years ago, and his book 'Elements' is still the ultimate geometry reference. He used construction techniques extensively. They give us a method of drawing things when simple measurement is not appropriate.

You should already know:

✔ how to draw and measure a line

✔ how to measure an angle

✔ how to use a pair of compasses to draw arcs and circles

✔ how to write ratios and reduce them to their simplest form

✔ how to convert between metric units

✔ how to recognise and draw enlargements of shapes

✔ the names of common quadrilaterals

✔ facts about common quadrilaterals

✔ the names of different types of triangle

✔ facts about right-angled triangles.

Key terms

arc
construction (construct)
equilateral triangle
perpendicular
bisector
vertex
scale
similar
congruent

Learn... 15.1 Drawing triangles accurately

There are different ways to draw a triangle accurately. The method you use depends on what you know about the triangle.

Drawing a triangle when all three sides are known

The following example shows how to draw a triangle when all three sides are known.

Example: Draw a triangle with sides 4.2 cm, 5.3 cm and 6 cm.

Solution: First draw a sketch to see what the triangle looks like.

This gives you an idea of what the finished triangle will look like and where to start on your page.

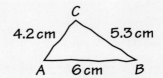

Draw and measure the line *AB* using a ruler and pencil. You should clearly mark end points on the line. Make sure you have enough space to draw the rest of the triangle.

Use your compasses to draw an **arc** of radius 5.3 cm with centre *B*.

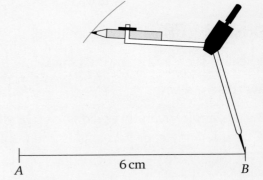

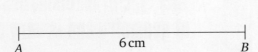

Use your compasses to draw an *arc* of radius 4.2 cm with centre *A*. Make sure your arcs cross each other.

Point *C* is where the two arcs cross. Join *C* to *A* and *C* to *B*. Label the sides with their lengths.

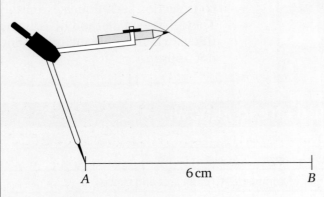

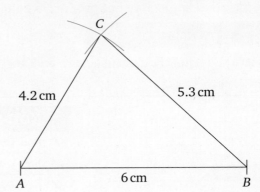

Drawing a triangle when one side and two angles are known

The following example shows how to draw a triangle when one side and two angles are known.

Example: Draw the following triangle accurately.

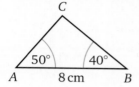

Solution: A diagram has been provided as part of the question, so there is no need to sketch a diagram.

Start by measuring and drawing the line *AB*.

8 cm

Use a protractor to measure an angle of 50° at point *A*. Mark the point and then join it to point *A*. Extend the line beyond the point you have marked. It is much better to have a line that is too long than too short.

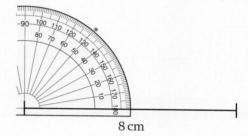

8 cm

> **AQA** *Examiner's tip*
>
> It is best to draw the two angles from either end of the given side. If one of these angles is the angle not given, then you can work out its size using the fact that angles of a triangle add up to 180°.

Repeat for the other angle (40°).

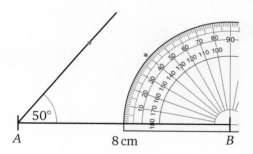

Now draw a line from *B*, so that angle *CBA* is 40°. Draw this line long enough so that it crosses the line drawn from *A*. Where these lines cross is point *C*.

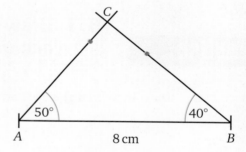

Label the triangle with two known angles and one known side.

Drawing a triangle when one angle and two sides are known

The following example shows how to draw a triangle when one angle and two sides are known.

Example: Draw the following triangle accurately.

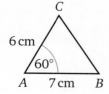

Solution:

Start with the longest side, *AB*.

Measure the angle given, using a protractor.

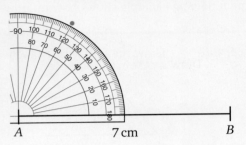

Draw in the side and remember to make it long enough. As before, it is better too long than too short.

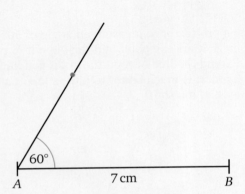

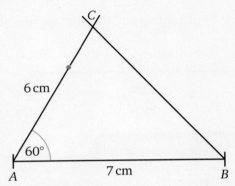

Now measure this side carefully.
Mark on the line the point where it should end 6 cm along.

Draw a line from this point to *B*. Label the triangle with the sides and angles you know.

15.1 Drawing triangles accurately

Practise...

G F E D C

1 **a** Draw these triangles accurately.

i 6 cm 8 cm 10 cm

ii 5 cm 8 cm 9 cm

iii 5 cm 7 cm 7 cm

b Measure the angles in your diagrams and check they add up to 180°.

2 Jack and Jill are asked to draw a triangle with sides of length 4 cm, 5 cm and 10 cm.
Jack says that he can draw this triangle. Jill says the triangle cannot be drawn.

Who is correct? Give a reason for your answer.

3 **a** Draw these triangles accurately.

i 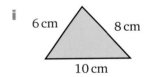 30° 30° 6 cm

ii 50° 10 cm 40°

iii 8 cm 25° 30°

b Measure the third angle in each triangle.
Check the angles in each triangle add up to 180°.

E

D

4   Draw this triangle accurately.

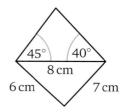

30°

8 cm

5   Draw this shape accurately.

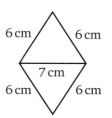

45°   40°

8 cm

6 cm   7 cm

6   Bill and Ben have been asked to draw this triangle accurately.

• The triangle has sides of 8 cm, 5 cm and a non-included angle of 30°.

They both draw it correctly, but their drawings look different.

Can you draw both Bill and Ben's diagrams?

7   Draw a rhombus accurately which has all sides equal to
6 cm and a shorter diagonal of 7 cm.

6 cm   6 cm

7 cm

6 cm   6 cm

 8   John is making a triangular prism from card.

Draw an accurate net for his prism.

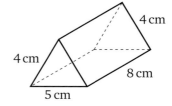

4 cm

4 cm

8 cm

5 cm

9   Jackie has a large packet of sweets. It weighs 1 kg. She is going to divide the sweets equally
between four of her friends. She wants to make four identical gift boxes from cardboard to contain
the sweets.

Design a suitable gift box that she could use for the sweets. Show clearly how she can make the gift
box from a flat sheet of card.

Learn...   **15.2 Constructions**   *k!*

Constructions are drawn using only a straight edge and a pair of compasses. You need to be able to
construct **equilateral triangles**, the **perpendicular** bisector of a line, and angle bisectors.

A **bisector** is a line which cuts something into two equal parts. A line bisector cuts a line into two equal
parts. An angle bisector cuts an angle into two equal parts.

Equilateral triangles

The following example shows how to construct an equilateral triangle.

Example: Construct an equilateral triangle.

Solution: Start with a line which will become one of the sides in your triangle, with a point *P*, where one **vertex** will be.

Open your compasses to the length of one side. With the point of your compasses on *P* draw a large arc that intersects the line at *Q*.

Keep the radius of your compasses the same. Put the point of the compasses on *Q* and draw an arc that passes through *P* and cuts the first arc at *R*.

Join *P* to *R*, and *Q* to *R*. You have now finished your construction.

You can use this technique to construct an angle of 60°. Follow the first two steps above, and then just join *P* to *R* (or *R* to *Q*).

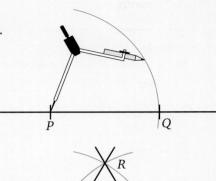

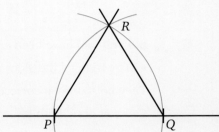

AQA Examiner's tip

Remember to leave your arcs. They show your method and will score you marks.

Line bisectors

The following example shows how to construct the bisector of a line.

Example: Construct the bisector of line *AB*.

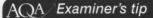

Solution: Open your compasses to more than half of *AB*. Put the point on *A* and draw arcs above and below *AB*.

Keep the radius of your compasses the same. Put the point of your compasses on *B* and draw two new arcs to cut the first two at *C* and *D*.

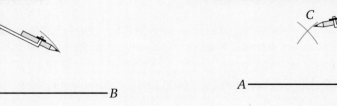

Join *CD*.

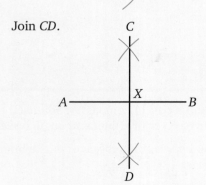

X is the midpoint of *AB*. *CD* not only bisects *AB*, it is called the perpendicular bisector of *AB*. This is because it meets *AB* at 90°.

Angle bisectors

The following example shows how to construct the bisector of an angle.

Example: Construct the bisector of angle *BAC*.

Solution:

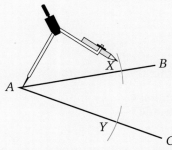
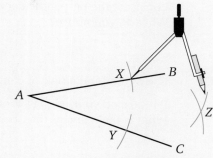

Open your compasses to less than the length of the shorter line. Put the point on *A* and draw arcs to cut *AB* at *X* and *AC* at *Y*.

Keep the radius of your compasses the same. Put the point of your compasses on *X* and *Y* in turn and draw arcs that intersect at *Z*.

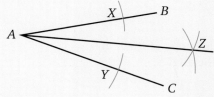

Bump up your grade

Learning to construct perpendicular bisectors and angle bisectors are Grade C skills.

Join *AZ*.
AZ is the angle bisector of angle *BAC*.

To construct an angle of 30° you first construct an angle of 60° (as part of an equilateral triangle) then bisect it.
To construct an angle of 45° you construct an angle of 90°, and then bisect it.
There are lots of other possible angles that can be constructed in a similar way.

Practise...   15.2 Constructions 🔧

G F E D C

1   Construct the perpendicular bisector of a line 8 cm long.

C

2   Draw this rectangle accurately.

Using only a ruler and compasses construct the perpendicular bisector of the diagonal *BD*.

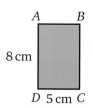

3   Construct an equilateral triangle of side 6 cm.

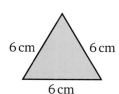

C

4 Draw a line 10 cm long.

    a Construct the perpendicular bisector of your line.

    b Your diagram shows four right angles. Bisect one of them to show an angle of 45°.

5 Accurately draw a triangle with sides 8 cm, 9 cm and 10 cm.
Construct the perpendicular bisector of each of the sides.
What do you notice?

6 Use a ruler and a pair of compasses to construct this net accurately.

Put tabs on your net, cut it out and make the shape.
What is the name of the 3D shape you have made?

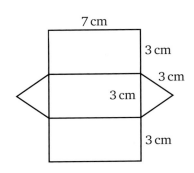

7 Accurately draw a triangle with sides 8 cm, 9 cm, and 10 cm.
Construct the angle bisector for each of the angles.
What do you notice?

8 Construct an angle of 60°. Bisect your angle to show an angle of 30°.

9 Harvey is asked to construct a perpendicular from a point, *P*, on a line, *l*.

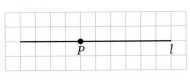

He starts by opening his compasses, and drawing arcs on *l* with the point of the compasses on *P*.

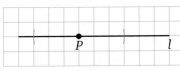

This gives him two points on *l* which are the same distance from *P*. He then opens his compasses a little more and draws arcs above *l*, as shown.

    a What does Harvey need to do next to complete his diagram?

    b Why did Harvey only draw arcs above *l*?

    c Copy and complete Harvey's diagram.

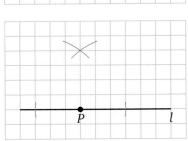

10   **a** Construct this rectangle.

    b Label your diagram carefully and draw in the diagonal *AC*.

    c Construct the perpendicular from vertex *B* to the diagonal *AC*.

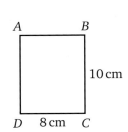

11 Harriet is asked to construct a perpendicular from a point, P, to a line, l.

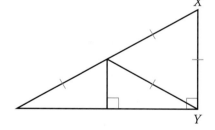

She opens her compasses and draws an arc crossing l with her compass point on P.

This gives Harriet two points on the line.
She now starts to construct the perpendicular bisector of the line between these points.

a Why does Harriet only need to draw arcs below the line for this last step?

b What does Harriet need to do to complete her diagram?

c Copy and complete Harriet's diagram.

12 Using a ruler and compasses only, construct a right-angled triangle which has a hypotenuse 10 cm long and a shorter side 8 cm long. Measure the third side. You may find it helpful to draw a sketch of the triangle first.

13 The diagram shows the roof design for a house.

Alan needs an accurate drawing of this roof to order wood from a supplier.

Construct the diagram $XY = 5$ cm.
Measure the width of the base of the roof to the nearest millimetre.

Learn... 15.3 Using scales

If something in real life is too large to draw you can use a **scale** to draw a smaller version.

Scales are used in models, plans and maps.

There are two ways to describe a scale:

- A scale of 1 centimetre to 1 kilometre means that 1 centimetre in the small version represents 1 kilometre in real life.

- A scale of 1 : 1000 means that every centimetre in the smaller version represents 1000 centimetres in real life.

Example:

a A plan has a scale of 1 centimetre to 5 kilometres. Write the scale as a ratio in its simplest form.

b A map has a scale of 1 : 200 000. What distance is represented by 1 centimetre on the map?

c The distance between two crossroads on a map is 12 centimetres. The distance between the crossroads in real life is 6 kilometres. What is the scale of the map?

d The scale on a map is 1 : 10 000. The distance between a church and a pub on the map is 5.3 centimetres. How far is it from the church to the pub in real life?

Solution:

a   1 centimetre on the plan represents 5 kilometres in real life.

First change the units so they are the same. Change the units from large (km) to small (cm).

5 kilometres = 5 × 1000 = 5000 metres (as there are 1000 metres in 1 kilometre).

5000 metres = 5000 × 100 = 500 000 centimetres (as there are 100 centimetres in 1 metre).

1 centimetre represents 500 000 centimetres in real life.

The scale is 1 : 500 000

b   1 : 200 000 means 1 centimetre on the map represents 200 000 centimetres in real life.

200 000 centimetres = 200 000 ÷ 100 = 2000 metres (as there are 100 centimetres in 1 metre).

2000 metres = 2000 ÷ 1000 = 2 kilometres (as there are 1000 metres in 1 kilometre).

1 centimetre represents 2 kilometres.

c   The information in the question tells you that 12 centimetres on the map represents 6 kilometres in real life.

Change the units from large (km) to small (cm).

6 kilometres = 6 × 1000 = 6000 metres (as there are 1000 metres in 1 kilometre).

6000 metres = 6000 × 100 = 600 000 centimetres (as there are 100 centimetres in 1 metre).

12 centimetres represent 600 000 centimetres.

The ratio is 12 : 600 000

This simplifies to 1 : 50 000 (by dividing both parts of the ratio by 12).

d   1 : 10 000 means 1 centimetre on the map represents 10 000 centimetres in real life.

5.3 centimetres on the map represents 5.3 times as far as 1 centimetre.

5.3 × 10 000 = 53 000 centimetres

53 000 centimetres is not a sensible choice of unit, so change to a more sensible one.

53 000 centimetres = 530 metres

Either 530 metres or 0.53 kilometres are sensible choices here.

Practise...   15.3 Using scales

E

1   How many kilometres do these lengths represent on a map that uses a scale of 1 : 70 000?

   a   8 cm   **b**   2.8 cm   **c**   8 mm

2   A map uses a scale of 1 : 25 000

Calculate the length on the map that represents a distance of:

   a   20 km   **b**   65.2 km   **c**   400 m.

3   A plan is drawn using a scale of 1 : 400

   a   A path on the plan has a length of 4.5 centimetres.
       How long is the actual path?

   b   A garden is 80 metres long in real life.
       What is the length of the garden on the plan?

4   A map has a scale of 1 : 20 000

   a   How long is a road which measures 7.5 centimetres on the map?

   b   A field is 400 metres long. How long is this field on the map?

5 A, B, C, are three points. B is 3 kilometres due south of C, and A is 4 kilometres due east of C.

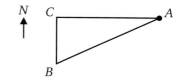

    a   Draw an accurate scale drawing showing the positions of A, B, and C. Use a scale of 1 : 50 000

    b   Measure the distance from A to B on the plan.

    c   What is the actual distance from A to B?

6   A ladder leans against a wall. The ladder reaches 5.5 metres up the wall and its base is 3.1 metres from the wall.

    a   Use a scale of 2 centimetres to 1 metre to make an accurate drawing showing the position of the ladder.

    b   Use your diagram to find the actual length of the ladder.

7   Molly draws a plan of her classroom using a scale of 1 : 40. The classroom is 8 metres long. Molly says her scale drawing will be 5 centimetres long. Is she correct? Give a reason for your answer.

8   Paul is a keen model maker. He makes the following models.

    a   **A Spitfire.** The scale of the model is 1 : 72 and the model is 12.2 centimetres long. How long is the actual Spitfire?

    b   **HMS Bounty.** The model is 37.2 centimetres long. The actual ship is 40.92 metres long. What is the scale of the model?

9   This is a sketch of Hilary's bedroom.

She plans to reorganise her room. She uses a scale drawing to help decide where to put her furniture.

She uses an A4 sheet of graph paper.

Draw an accurate plan of her bedroom using an appropriate scale.

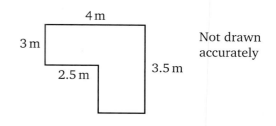

10   The diagram shows the front elevation of a garage.

The manufacturer claims the height of the garage at its highest point is 4.4 metres.
Is this claim correct?

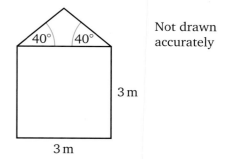

11   This diagram shows the space under the stairs in a house.

Jack wants to put a cupboard under the stairs. The cupboard is 1.6 metres high, 60 centimetres deep and 1.2 metres wide. It has two doors on the front, each is 60 centimetres wide.

Does this cupboard fit under the stairs? If so, can the doors open? Show working to justify your answers.

Scale: 1 cm = 0.2 m (20 cm)

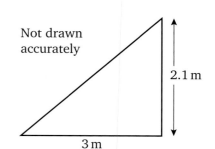

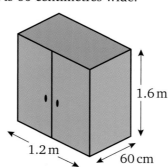

Learn...   15.4 Similarity and congruence

Two shapes are mathematically **similar** if they have the same shape but different sizes. So one shape is an enlargement of the other with an associated scale factor.

Two shapes are **congruent** if they have both the same shape and the same size. That is, the two shapes are identical. If you cut one of the shapes out then it will fit on top of the other. You may need to turn it around or 'flip' it over to do this.

Link

See Chapter 14 for more on enlargements and scale factors.

Example: Look at the shapes in the diagrams below.

The two shapes labelled **a** and **b** are **similar**. They have the same shape but different sizes. Shape **b** is an enlargement of shape **a**. Here the scale factor is two, since all the lengths in shape **b** are twice as long as the lengths in shape **a**.

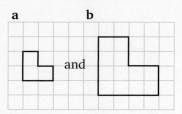

The two shapes labelled **c** and **d** are **congruent**. They have the same shape and size. They are identical.

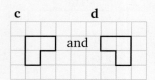

The two shapes labelled **e** and **f** are neither congruent nor similar. They are different shapes. Although it may look as though shape **f** is an enlargement of shape **e**, in fact the lengths have not all been increased by the same factor.

AQA *Examiner's tip*

In an enlargement all the sides are multiplied by the same factor.

Practise...   15.4 Similarity and congruence   G F E D C

G

1   Name the pairs of congruent shapes in this diagram.

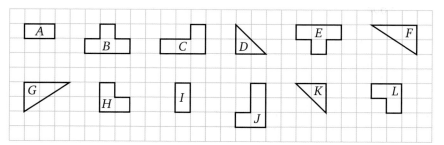

2

Which of the following shapes are similar to the shape above?

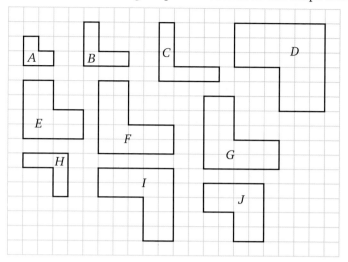

3   Group these shapes into sets of congruent shapes.

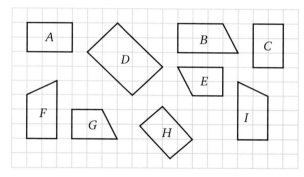

4   Which of the following shapes are similar to the red triangle?

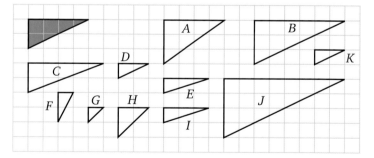

5   Group these shapes into sets of similar shapes.

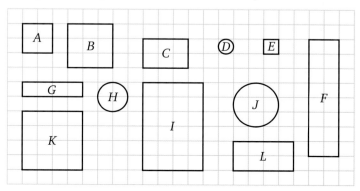

F

G
F

F

F

6   **a**   Tim says, 'Any two squares are similar to each other.' Is Tim correct?
Give a reason for your answer.

b   Jim says, 'Any two rectangles are similar to each other.' Is Jim correct?
Give a reason for your answer.

c   Pam says, 'Any two circles are similar to each other.' Is Pam correct?
Give a reason for your answer.

d   Sam says, 'Any two triangles are similar to each other.' Is Sam correct?
Give a reason for your answer.

?   **7**   Join the dots on a copy of the grid to make a triangle.

a   Find as many **different** triangles as you can. Similar
triangles are allowed, but not congruent triangles.
How can you check you have found all the possible
triangles?

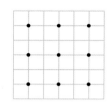

Write down as many facts about each triangle as you can.
Sort them into groups which have facts in common.

b   Using the same 3 × 3 dotted grid, find as many different quadrilaterals as you
can. Similar quadrilaterals are allowed, but not congruent quadrilaterals.
How can you check you have found them all?

Write down as many facts about each quadrilateral as you can.
Sort them into groups which have facts in common.

15   Assess   k!

F

1   Which of these shapes are similar and which are congruent?

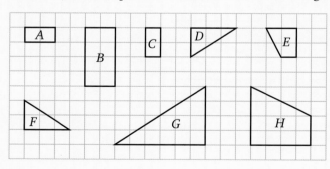

E

2   Draw these triangles accurately.

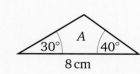 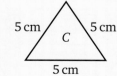 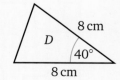

Not drawn
accurately

3   The scale on a map is 1 : 50 000. How many kilometres do each of these lengths
on the map represent in real life?

  a   8 cm   **b**   2.8 cm   **c**   8 mm

4 A map uses a scale of 1:25 000. Calculate the length on the map that represents a distance of:

    a   10 km   **b**   24.8 km.

E

5 A maths class is asked to draw this triangle accurately.

Explain why they may not all agree on the answer.

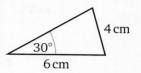

D

6 Draw this kite accurately.

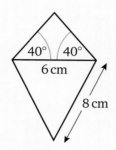

7   **a** Draw this rectangle accurately.

      b Bisect angle *ABC*.

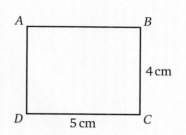

Not drawn accurately

C

8   **a** Draw this diagram accurately.

      b Construct the perpendicular bisector of *AB*.

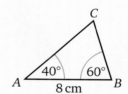

Not drawn accurately

9 This diagram shows the sails on Rob's boat.

Draw an accurate diagram of the sail.

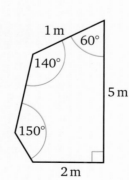

Not drawn accurately

AQA Examination-style questions

1 The side of a rhombus is 7 cm.
The length of the shorter diagonal is 5 cm.

Make an accurate drawing of the rhombus.

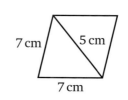

Not drawn accurately

(4 marks)

AQA 2007

16 Loci

Objectives

Examiners would normally expect students who get these grades to be able to:

G

measure and draw lines accurately

F

measure and draw angles accurately

E

use map scales to find a distance

D

understand the idea of a locus

C

construct the locus of points equidistant from two fixed points

construct the locus of points equidistant from two fixed lines

solve loci problems, for example the locus of points less than 3 cm from a point *P*.

Key terms

locus, loci
perpendicular
bisect, bisector
equidistant

Did you know?

Mobile phone masts

Some mobile phone masts have a range of 40 km.
Hills, trees and buildings can all reduce this distance to as little as 5 km.
In some places, mobile phone masts are only 1 or 2 km apart.
This is because they could not cope with the number of calls being made in the area on their own.

You should already know:

✔ how to measure a line accurately
✔ how to measure and draw an angle accurately
✔ how to construct the perpendicular bisector of a line
✔ how to construct the bisector of an angle
✔ how to construct and interpret a scale drawing.

16.1 Drawing and measuring lines, angles and circles

When drawing or measuring a line, take care to be as accurate as possible.

Start measuring from 0 on the ruler. This line is 7.6 cm long.

When drawing or measuring an angle, put the central mark on the vertex.

Make sure you use the scale that starts at 0. This angle is 103°. The inner scale starts at 0.

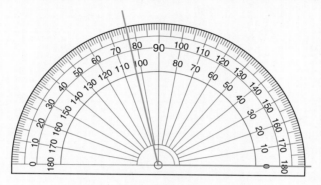

To draw a circle, measure the distance between your compass point and pencil point carefully against a ruler.

Example: The diagram shows a parallelogram *ABCD*, with *CD* = *BA* = 8 cm, *AD* = *BC* = 6 cm

Angle *ADC* = 68° and *DAB* = 112°

A semicircle is drawn on *AB*.

Make an accurate drawing of this figure.

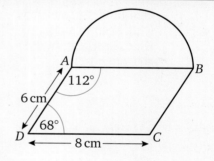

Solution: Draw and label a line, *DC*, 8 cm long.

Angle *ADC* is the angle between *AD* and *DC*, so it is at *D*.

At *D*, draw an angle of 68° from *DC*.

Measure 6 cm along this line, and label the point *A*.

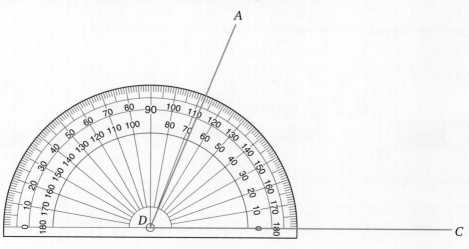

At *A*, draw an angle of 112° from *DA*.

Measure 8 cm along this line, and label the point *B*.

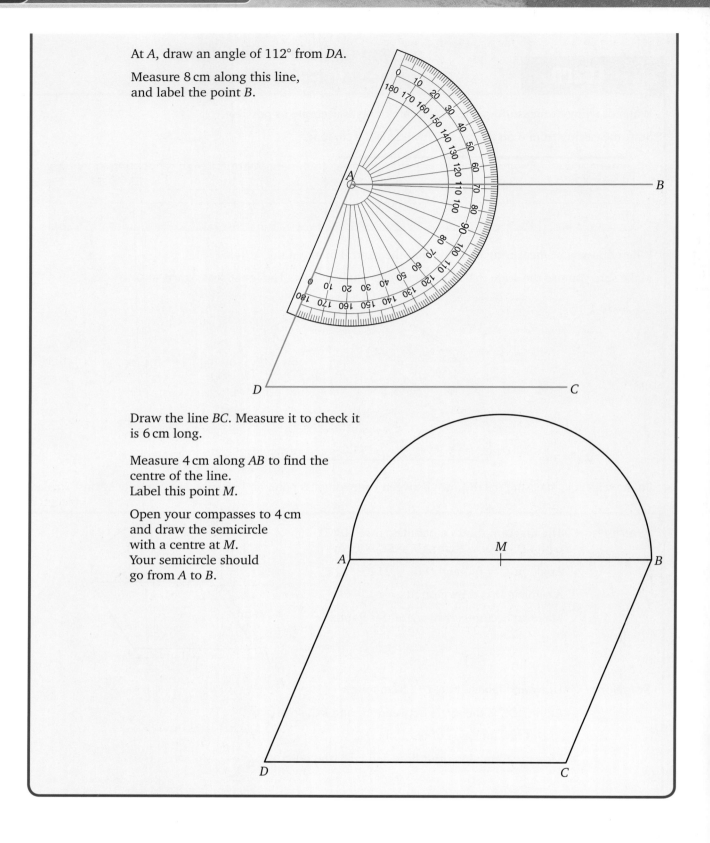

Draw the line *BC*. Measure it to check it is 6 cm long.

Measure 4 cm along *AB* to find the centre of the line. Label this point *M*.

Open your compasses to 4 cm and draw the semicircle with a centre at *M*. Your semicircle should go from *A* to *B*.

16.1 Drawing and measuring lines, angles and circles

Practise...

G

1 Draw a rectangle 8.2 cm long and 4.6 cm wide.

a Draw a diagonal of the rectangle.

b Measure and write down the length of the diagonal.

2   Triangle ABC has side AB = 7.4 cm, and BC = 5.2 cm
Angle ABC (the angle at B) is a right angle.

    a   Make an accurate drawing of the triangle.

    b   Measure and write down the length of AC.

    c   Measure and write down the size of angles BAC and ACB.

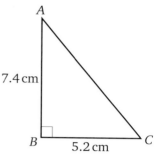

3   In trapezium ABCD, AB and DC are parallel.
Make an accurate drawing of the trapezium.

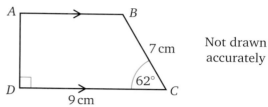

Not drawn accurately

4   Triangle XYZ has XY = 7.2 cm, YZ = 8.4 cm and angle
XYZ = 54° as shown.

Make an accurate drawing of the triangle.

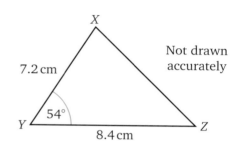

Not drawn accurately

5   A rhombus has sides of 8 cm and angles of 72° and 108°.

    a   Make an accurate drawing of the rhombus.

    b   Draw a circle that just touches the four sides
of the rhombus.
Show clearly how you found the centre
of the circle.

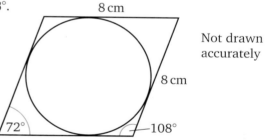

Not drawn accurately

6   A hockey pitch has these markings and dimensions.

Make a scale drawing of the pitch, using 1 cm to represent 10 m.

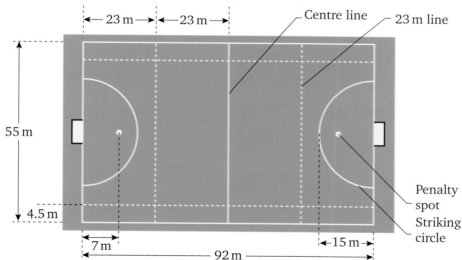

7   **a**   Draw a rectangle, ABCD, with AB = 8.2 cm and AD = 5.7 cm

    b   A circle passes through A, B, C and D.
Draw this circle, showing clearly on your diagram
how you found its centre.

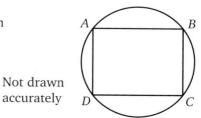

Not drawn accurately

Learn... 16.2 Describing a locus

A **locus** can be thought of in two different ways.

It is the path that a moving point follows, or a set of points that follow a rule.

For example, a circle with a radius of 10 cm, centre C, can be thought of as all the points 10 cm from C, or as the path of a moving point which is always 10 cm from a fixed point, C.

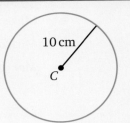

Example: Find the locus of a point on a train moving on a straight track.

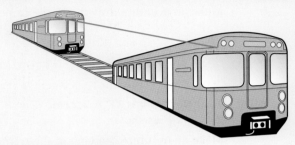

Solution: It follows a path that is a straight-line segment, shown here in red.

Example: Sketch the locus of the vertex A of the triangle ABC, as it rotates clockwise about C until CB is horizontal.

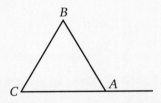

Solution: The diagram shows how the point A (marked in red) moves in an arc (or part of a circle.)

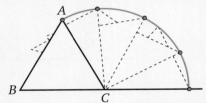

AQA *Examiner's tip*

As well as visualising the locus, always look for any fixed points in the situation, in this example, point C.

Practise... 16.2 Describing a locus 🔵 G F E D C

G

1 Alan, Nicky and Margaret are in a PE lesson.

They have to run from a starting point, S, to touch the wall, and then run to the finishing point F.

The diagram shows their paths.

Use the diagram to find out who runs the shortest distance.

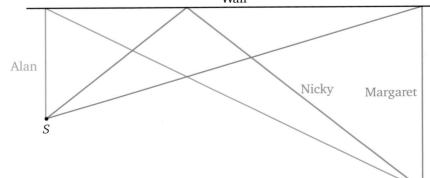

2 Gill goes to a park to play.

She goes on two slides, a swing and a roundabout.

a Sketch the side view of the locus of her head as she goes down the slides.

b Sketch the side view of the locus of her head as she plays on the swing.

c Sketch the plan view of the locus of her head as she rides on the roundabout when viewed from above.

3 Mark a point, X, on a piece of paper.

Place some counters so that their centres are exactly 5 cm from the point.

Draw the shape that your counters would make if you had an unlimited supply.

4 Draw a line 10 cm long.

Place some counters so that their centres are exactly 5 cm from the line.

Draw the shape that your counters would make if you had an unlimited supply.

5 Bill throws a ball from his window.

Harry, Hope and Oli try to sketch the locus of the ball.

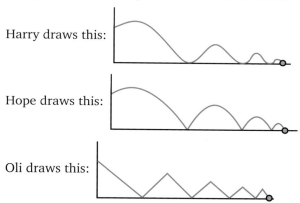

Harry draws this:

Hope draws this:

Oli draws this:

Who is correct?

Give a reason for your answer.

6 Most cars have one of the three arrangements of windscreen wipers shown below.

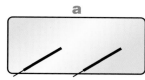 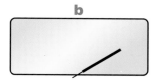 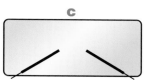

For each arrangement, sketch the area of windscreen that the wipers will clear.

Which is the best arrangement? Give a reason for your answer.

7 A square is rolled along a straight line. Sketch the locus of a vertex of the square as it is rolled.

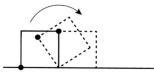

Check your answer by cutting a square from card, and rolling it along a ruler.

D

8 On a separate diagram, sketch the locus of:

a the end of the minute hand of the clock

b the end of the pendulum on a grandfather clock

c a mouse running up the pendulum of the clock.

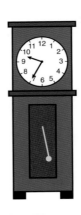

9 Sketch the locus of a piece of chewing gum stuck to a car wheel as the car moves.

Hint

Mark a point on the edge of a coin and roll it along a ruler (without letting it slip) to see how the point moves.

10 Describe the locus of:

a all the points 5 cm from a fixed point, *P*

b all the points 2 cm from the circumference of a circle of radius 6 cm.

11 Describe the loci shown with red lines in the diagrams below.

a

b

c

d

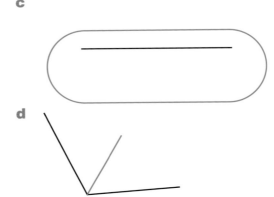

12 Two cars are waiting at a roundabout.

Car *A* is going to take the first exit (the one on the left).

Car *B* is going to take the third exit (the one on the right).

Draw the loci of the two cars.

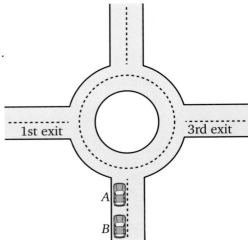

1st exit   3rd exit

A
B

13 Place two 5p coins together as shown.

a Hold one still and roll the other one all the way round it.
How many times does the coin rotate as it goes round the stationary one?
Use two coins, counters or discs to find the answer.

b Find the locus of a point on the moving coin as it goes round.
What would happen if one coin had a diameter twice the size of the other?

 **Learn...**   **16.3 Constructing loci**

You need to remember your constructions from Chapter 15.

It is always useful to draw a sketch first before tackling any locus question.

A **perpendicular bisector** of the line *AB* joins all the points **equidistant** from (the same distance from) *A* and *B*.

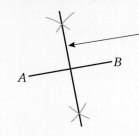

Locus of points equidistant from *A* and *B*: perpendicular bisector

Draw four arcs of the same radius, two with centre at *A*, two with centre at *B*. Join the points of intersection.

An angle bisector joins all points equidistant from two lines that meet at the angle.

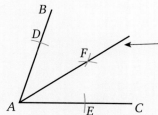

Locus of points equidistant from *AB* and *AC*: angle bisector

Draw two arcs of equal radius, centre *A*, to cut *AB* and *AC* at *D* and *E*. Draw equal arcs from *D* and *E* to intersect at *F*.

This circle joins all the points that are 1 cm from point *A*.

Inside the circle are all points less than 1 cm from *A*.

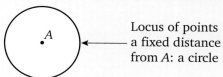

Locus of points a fixed distance from *A*: a circle

The circle is made up of all points 1 cm from *A*.

AQA Examiner's tip

The locus of a point equidistant from two fixed points and the locus of points equidistant from two fixed lines are the only constructions you need to know.

The locus of all points 1 cm from a line *AB* is shown below.

The parallel lines are joined by semicircles.

All the points less than 1 cm from *AB* are inside the closed shape.

All the points more than 1 cm from *AB* are outside the closed shape.

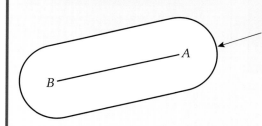

Locus of points a fixed distance from *AB*: parallel lines with semicircular ends

AQA Examiner's tip

To get full marks, make sure you leave your construction lines showing.

Example: An electricity pylon has to be placed so that it is equidistant from *AB* and *AC*, and no more than 200 m from *D*.
It must be within the boundary.

Mark the points where the pylon could be placed.

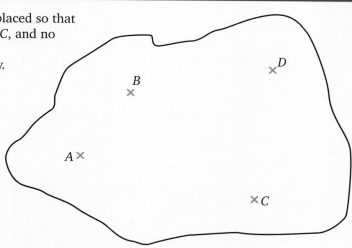

Scale: 1 cm = 100 m

Solution: Draw *AB* and *AC*.

Construct the angle bisector of *BAC*, as this marks the locus of points equidistant (the same distance) from *AB* and *AC*.

The points less than 200 m from *D* form a circle, radius 200 m.

With a scale of 1 cm to 100 m, this circle needs to have a radius of 2 cm.

The possible positions for the pylon are on the angle bisector and inside the circle and the boundary.

This is shown in green.

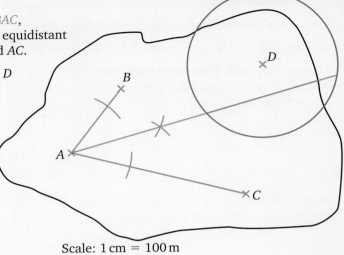

Scale: 1 cm = 100 m

Practise...   16.3 Constructing loci   *k!*   G F E D C

1   Alice, Kat and Becky tried to draw the locus of points a distance of 1 cm outside a rectangle *ABCD*.

Here are their answers.

Alice   Kat   Becky

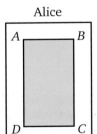 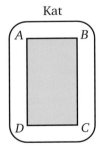 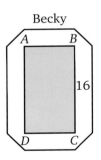

Who is correct: Alice, Kat or Becky?

Give a reason for your answer.

2   **a**   Draw a line, *AB*, 8 cm long.

      b   Use a ruler and compasses to construct the locus of points equidistant from *A* and *B*.

3   **a**   Draw an angle, *ABC*, of 70°.

      b   Use a ruler and compasses to construct the locus of points equidistant from *AB* and *BC*.

4   Draw a line, 8 cm long. Label it *AB*.

      a   Find the locus of points that are 6 cm from *A*.

      b   Find the locus of points that are 4 cm from *B*.

      c   Shade the area containing all the points that are less than 6 cm from *A* and less than 4 cm from *B*.

5 Before answering this question, draw a sketch first so you can be sure it will fit on your page.

ABC is a triangle.

Angle *ABC* = 90°, *AB* = 8 cm and *BC* = 6 cm

  a Make an accurate drawing of triangle *ABC*.

  b Use a ruler and compasses to construct the locus of points equidistant from *A* and *C*.

  c Construct the locus of points 6 cm from *C*.

  d Mark the points that are equidistant from *A* and *C* and are also less than 6 cm from *C*.

6 Draw a triangle *ABC* with sides at least 10 cm long.

  a Use a ruler and compasses to construct the locus of points equidistant from *A* and *B*.

  b Use a ruler and compasses to construct the locus of points equidistant from *A* and *C*.

  c Mark the point that is equidistant from *A*, *B* and *C*. Label it *X*.

7 Draw a triangle *ABC* with sides at least 10 cm long.

  a Draw the locus of points equidistant from *AB* and *BC*.

  b Draw the locus of points equidistant from *AB* and *AC*.

  c Mark the point that is equidistant from *AB*, *BC* and *AC*. Label it *X*.

8 Tommy has a rectangular garden, *ABCD*, 12 m long by 8 m wide.

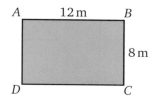

He wants to plant a tree in the garden.

He wants the tree to be at least 3 m from the edge *CD* of the garden.

It must be no more than 6 m from *B*.

Using a scale of 1 cm to 1 m, make a drawing of the garden, and shade the region where Tommy can plant the tree.

9 In Suffolk, there are two mobile phone masts, 40 km apart.

  a Make a map showing the two masts, using a scale of 1 cm to 4 km. Label them *A* and *B*.

  b Each mast has a range of 32 km if you are using a phone outside a building.

  If you use a phone inside a building the range drops to 24 km.

   i Shade the area where you can get a signal if you are outside.

   ii Shade the area where you can get a signal inside a house.

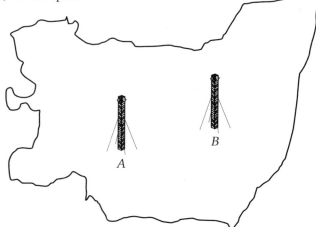

10 A rectangular lawn is 12 m long and 10 m wide.

A gardener waters the lawn with sprinklers, which spray water in a circle with a radius of 3 m.

Find the smallest number of sprinklers needed to water the entire lawn.

16   Assess   k!

F

1   Measure the sides and angles of this triangle.

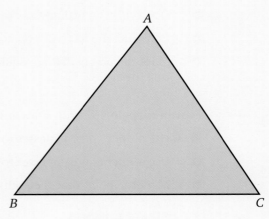

D

2   Draw a parallelogram, *ABCD*, with *AB* = 7.5 cm, *BC* = 6.8 cm and angle *ABC* = 66°

3   Make an accurate copy of the shape drawn below.

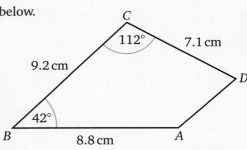

Not drawn accurately

4   Sketch the locus of point *A* on the key as it unlocks a door.

C

5   The diagram shows the plan, *ABCD*, of a garden. The scale is 1 cm to 2 m.

    a   A gnome, *G*, is to be placed 12 m from *C* and equidistant from *AB* and *AD*.

        On a copy of the diagram, construct and mark the position of *G*.

    b   Find the real distance of the gnome from *D*.

6   A triangle *ABC* has *AB* = 8 cm, *BC* = 6.5 cm and angle *ABC* = 72°.

    a   Make an accurate drawing of the triangle.

    b   Use a ruler and compasses to construct the points that are the same distance from *BC* and *AC*.

7   Mark two points, *A* and *B*, 9 cm apart.

    Shade the region that is no more than 6 cm from *A* and no more than 5 cm from *B*.

AQA Examination-style questions

1   *AB* and *AC* represents two walls.
A mast is to be erected that is:

- equidistant from *AB* and AC
- between 40 m and 70 m from *A*.

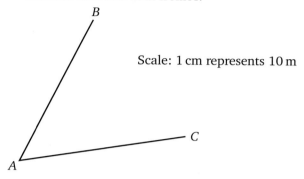

Scale: 1 cm represents 10 m

Show clearly all the possible positions of the mast.

(3 marks)

AQA 2006

2   Using a ruler and compasses construct the bisector of angle *ABC*.

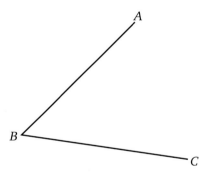

(2 marks)

AQA 2007

17 Quadratics

Objectives

Examiners would normally expect students who get these grades to be able to:

D

draw graphs of simple quadratics such as $y = x^2$, $y = x^2 - 4$ and $y = 3x^2$

C

draw graphs of harder quadratics such as $y = x^2 + 2x + 1$

use a quadratic graph to estimate x- and y-values, giving answers to an appropriate degree of accuracy.

Did you know?

Quadratic equations in action

Why are car headlights so bright?

The reflecting surface inside a car headlight is curved and the equation of its surface fits a quadratic equation.

When the light from the bulb is reflected from the surface, the light is brighter because all the rays are projected forward in a concentrated beam.

Key terms

quadratic expression
quadratic equation
parabola
symmetrical
line of symmetry

You should already know:

✔ how to plot coordinates in all four quadrants

✔ how to draw straight-line graphs

✔ how to substitute positive and negative numbers into an expression.

Learn... 17.1 Drawing graphs of simple quadratics

A **quadratic expression** has x^2 as its highest power.

x^2, $x^2 + 3$, $2x^2 - 3x + 4$ and $3x^2 + 5x - 2$ are all examples of quadratic expressions.

The following expressions are **not** quadratics: $x + 7$, $x^3 - 3x^2$ and $x^2 + x^4$

How to draw the graph of a quadratic

1. **Complete or construct a table of values.**

   This is the table for the **quadratic equation** $y = x^2$

x	−3	−2	−1	0	1	2	3
y	9	4	1	0	1	4	9

   When drawing the graphs of straight lines (linear graphs), you were taught to plot three points.

   The graphs of quadratics are all curves, so you need more than three values in the table.

   The y-values are found by substituting the x-values into the equation of the quadratic, e.g. for $y = x^2$

   when $x = 2$ $y = x^2 = 2^2 = 4$

   when $x = -3$ $y = x^2(-3)^2 = 9$

   Sometimes you are asked to construct the table for yourself.

   Make sure that you construct this table for the range of values given in the question.

2. **By looking at your table, find the smallest and largest y-values that you will need on the y-axis.**

3. **Draw a pair of axes for your graph and label them x and y.**

4. **From your table of values, plot the points on the graph as small crosses.**

   The points from this table would be $(-3, 9)$, $(-2, 4)$, $(-1, 1)$, $(0, 0)$, $(1, 1)$, $(2, 4)$, $(3, 9)$

5. **Join the points with a smooth curve.**

Graph of $y = x^2$

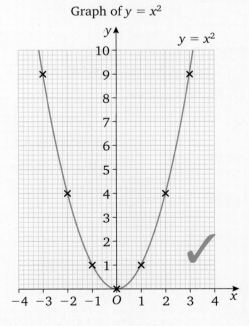

NOT

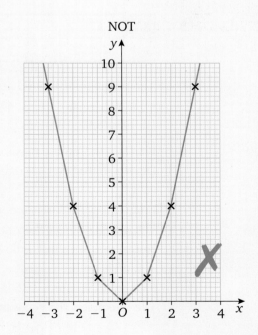

It should be smooth at the bottom.

Always draw your line with a sharp pencil, making sure that it passes clearly through the centre of the points plotted.

The graph obtained is a curve. This shaped curve is called a **parabola**.

$y = x^2$ is a U-shaped graph and has a minimum (lowest) point at (0, 0).

Note that the x^2 term is **positive** for a **U-shaped** curve.

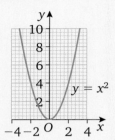

$y = x^2$

Other quadratics have maximum (highest) points and are hill-shaped graphs.

Note that the x^2 term is **negative** for a **hill-shaped** curve.

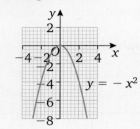

$y = -x^2$

All quadratics are parabolas. They are also **symmetrical**, having one **line of symmetry**.

You can sometimes see the pattern in the table of x- and y-values.

Example:

a Copy and complete the table of values for the quadratic $y = x^2 + 3$

x	−3	−2	−1	0	1	2	3
y		7	4		4		12

b Draw the graph for the values of x shown in the table.

c From your graph, find the value of y when $x = 1.5$

d From your graph, find the values of x when $y = 7$

Solution:

a When $x = -3, y = (-3)^2 + 3 = 12$
When $x = 0, y = 0^2 + 3 = 3$
When $x = 2, y = (2)^2 + 3 = 7$

x	−3	−2	−1	0	1	2	3
y	12	7	4	3	4	7	12

Notice that the graph is **symmetrical** about the y-axis.

You can see the symmetry in the table. The values for $x = -3$ and $x = 3$ give the same y-values.

b The points to be plotted are:
(−3, 12), (−2, 7), (−1, 4), (0, 3), (1, 4), (2, 7) and (3, 12)

c Find $x = 1.5$ on the x-axis.
Draw a dotted vertical line up to meet the curve.
From this point, draw a dotted horizontal line across to meet the y-axis.
Notice that with this scale 10 small squares = 2 units on the y-axis.
So 1 small square = 0.2
The line meets the y-axis at $y = 5.2$

Graph of $y = x^2 + 3$

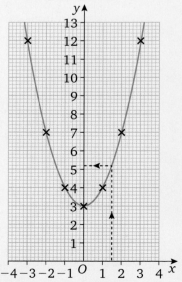

AQA *Examiner's tip*

If you are right-handed, you may find it easier to join the points by turning your page upside-down.

AQA *Examiner's tip*

Make sure the curve passes through the middle of all the plotted points. Join the points with one continuous curved line. If you join them with straight lines, you will lose marks.

d Find $y = 7$ on the y-axis.

Draw a dotted horizontal line to the left and the right to meet the curve in two places.

From each of these points, draw a dotted vertical line down to meet the x-axis.

These lines meet the x-axis at $x = -2$ and $x = 2$

With harder questions you can use an alternative method which has extra working in the table as shown in the second example below.

Graph of $y = x^2 + 3$

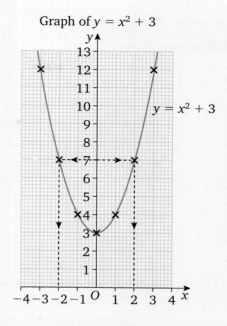

Example:

a Draw the graph of $y = x^2 + 2x$ for values of x from -3 to 2. Use the axes provided.

b Identify the line of symmetry on your graph.

c Write down the coordinates of the lowest (minimum) point.

Solution:

a

Add these rows to get the y-values

x	-3	-2	-1	0	1	2
x^2	9	4	1	0	1	4
$+2x$	-6	-4	-2	0	2	4
y	3	0	-1	0	3	8

The coordinates of the points to be plotted are found in the top row (x) and the bottom row (y).

For example, $(-3, 3)$, $(-2, 0)$, $(-1, -1)$

b This graph is symmetrical about a vertical line through $x = -1$

c The minimum point is the lowest point on the graph, $(-1, -1)$. At this point, y has its smallest value.

Graph of $y = x^2 + 2x$

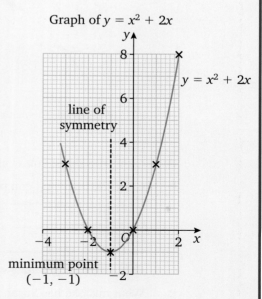

line of symmetry

minimum point $(-1, -1)$

AQA Examiner's tip

All quadratic graphs should be smooth curves.

They are either U-shaped or hill-shaped.

If your graph is not either of these, check your working in the table.

A quadratic curve never has a flat base.

17.1 Drawing graphs of simple quadratics

Practise...

G F E D C

D

1 **a** Copy and complete the table of values for $y = x^2 + 1$

x	−3	−2	−1	0	1	2	3
y	10			1			10

b Draw the graph of $y = x^2 + 1$ for values of x from −3 to 3.
(You will need y-values from 0 to 10.)

2 **a** Copy and complete the table of values for $y = 2x^2$

x	−3	−2	−1	0	1	2	3
y		8		0	2		18

Hint

Remember that $2x^2$ means 2 times x^2. So square first then multiply the answer by 2.

b Draw the graph of $y = 2x^2$ for values of x from −3 to 3.
(You will need y-values from 0 to 18. Label the y-axis 0, 2, 4, …, 18.)

c Write down the coordinates of the minimum point on the graph.

3 The graph shows the points that Susan has plotted for her graph of $y = 3x^2$

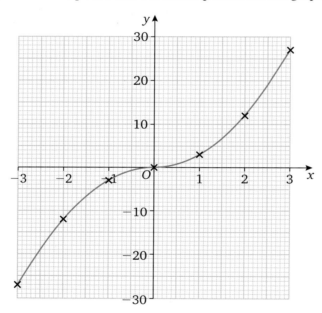

a Is this the shape you would expect? Give a reason for your answer.

b Copy the table and complete it for $y = 3x^2$.

x	−3	−2	−1	0	1	2	3
y							

Compare your y-values with those for Susan's graph.

c What mistake has Susan made in calculating the values?

4 **a** Draw the graph of $y = -x^2$ for values of x from −3 to 3.

b This graph has **no** minimum point. It has a maximum point.
How can you tell this from the equation?

c Write down the coordinates of the maximum point.

5 Here is the table for drawing the graph of $y = x^2 - 2x$

x	-3	-2	-1	0	1	2	3
x^2					1	4	9
$-2x$	6	4			-2		
y					-1		

} Add

> **Hint**
>
> You add the two middle rows to get the y-values.
>
> Be careful with the signs.

a Compare this with the table in the second example in Learn 17.1. What is the difference between the two tables?

b Copy and complete this table.

c Use your table to draw the graph of $y = x^2 - 2x$ for values of x from -3 to 3.

d What is the minimum point on this graph?

 6

a Construct a table for $y = x^2 - x$ for values of x from -2 to 4.

b Plot the points from the table and join them up to make a smooth curve.

c By looking at the symmetry of the graph, what is the x-coordinate of the minimum point?

d Calculate the corresponding y-value and state the coordinates of the minimum point.

7

a Construct a table for $y = -x^2 + 2x$ taking values of x from -2 to 4.

b Plot the points from the table and join them up to make a smooth curve.

c Find the coordinates of the maximum point on the graph.

d From your graph, find the value of y when $x = 3.5$

e From your graph, find the values of x when $y = 2$

> **Bump up your grade**
>
> Grade C students should be able to read off corresponding x- and y-values from a quadratic graph and give their answers to an appropriate degree of accuracy.

8 The quadratic equation $d = \dfrac{s^2}{200} + 1$ gives the stopping distance, d (in car lengths), of a car moving at s mph.

According to this formula, at 20 mph the stopping distance is three car lengths.

When $s = 20$, $d = \dfrac{20^2}{200} + 1 = 3$

According to this formula, at 40 mph the stopping distance is nine car lengths.

When $s = 40$, $d = \dfrac{40^2}{200} + 1 = 9$

Copy and complete the table for the speeds shown.

s	20	30	40	50	60	70	80
d	3		9			25.5	

From this table, draw the graph showing how the stopping distance relates to the speed.

 9

a Taking values of x from -4 to 4, draw tables of values for the quadratics $y = x^2$, $y = x^2 + 3$ and $y = x^2 - 4$

b On the same grid and using the same axes and scales, draw the graphs of these quadratics.

c Compare your graphs. What effect does changing the number at the end of the quadratic have on the graph?

Learn... 17.2 Drawing graphs of harder quadratics

The quadratics already drawn have consisted of one or two terms.

For example, $y = \underline{x^2}$ and $y = \underline{x^2} + \underline{4x}$
'1 term'   '2 terms'

For the exam you may be asked to draw the graph of a quadratic that contains three terms.

For example, $y = \underline{x^2} + \underline{2x} + \underline{3}$ or $y = \underline{5} + \underline{x} - \underline{x^2}$
'3 terms'   '3 terms'

You may be asked to find the value of y for a given value of x or the value of x for a given value of y.

> **Bump up your grade**
>
> If you can construct tables for equations consisting of three terms, as in $y = x^2 + 2x + 3$ or $y = 5 + x - x^2$ and draw their graphs, you will be working at Grade C.

Example:

a Copy and complete the table for $y = 6 + x - x^2$

x	−3	−2	−1	0	1	2	3	4
y	−6	0			6			

b What are the coordinates of the maximum point?

c Using your table, draw the graph of $y = 6 + x - x^2$ for $-3 \leqslant x \leqslant 4$

d What is the value of y when:

i $x = 2.8$ **ii** $x = -1.2$?

Give your answers to 1 d.p.

> **Hint**
>
> For $-3 \leqslant x \leqslant 4$ means that you take x-values from −3 to 4 inclusive.

Solution:

a This table has only two rows, x and y. As some of the values are already given, complete the table by substitution.

So when

$x = -1$, $y = 6 + x - x^2 = 6 + (-1) - (-1)^2 = 6 - 1 - 1 = 4$

$x = 0$, $y = 6 + 0 - 0^2 = 6$

$x = 2$, $y = 6 + 2 - 2^2 = 4$

$x = 3$, $y = 6 + 3 - 3^2 = 0$

$x = 4$, $y = 6 + 4 - 4^2 = -6$

x	−3	−2	−1	0	1	2	3	4
y	−6	0	4	6	6	4	0	−6

If you were forming the table yourself you could use either the extended table or the shortened form as used here.

The extended table would begin like this.

All the entries are the same on this line. → The term does not contain an x term so the answers do not change.

x	−3	−2	−1	0	1	2	3	4
6	6	6	6					
+x	−3	−2						
−x^2	−9	−4						
y	−6	0						

The quadratic contained three terms. Add these three rows.

The sign is included with the term

b The table shows that the graph is symmetrical about a point halfway between $x = 0$ and $x = 1$. This means that there will be a maximum point when $x = \frac{1}{2}$ (or 0.5)

At $x = 0$ and $x = 1$, the y-value is 6, so the y-value of the maximum point will be slightly larger than 6.

To get an accurate value, the y-value is found by substituting into the equation.

When $x = 0.5$, $y = 6 + 0.5 - (0.5)^2 = 6 + 0.5 - 0.25 = 6.25$

There is a maximum at $(0.5, 6.25)$.

c   Graph of $y = 6 + x - x^2$

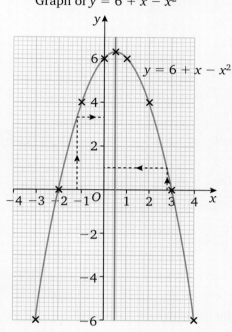

d i   When $x = 2.8$, $y = 1.0$

ii   When $x = -1.2$, $y = 3.3$

Bump up your grade

Grade C students should be able to find the solutions of quadratic equations such as $x^2 + 2x - 8 = 0$ and understand why these are found from reading off values on the x-axis.

17.2 Drawing graphs of harder quadratics

Practise...

G F E D C

C

1 a Copy and complete this table for $y = x^2 - 2x - 8$

x	-2	-1	0	1	2	3	4	5
x^2	4	1	0		4		16	25
$-2x$	$+4$		0	-2	-4	-6	-8	-10
-8	-8		-8		-8			-8
y	0		-8		-8			7

b Draw the graph of $y = x^2 - 2x - 8$ for values of x from -2 to 5.

c When $x = 3$, what is the value of y?

d Work out the values of x for $y = 40$.

C

2   Julie constructed a table for $y = x^2 - x - 6$ for values of x from -2 to 4.

x	-2	-1	0	1	2	3	4
x^2	4	1	0	1	4	9	16
$-x$	-2	-1	0	-1	-2	-3	-4
-6	-6	-6	-6	-6	-6	-6	-6
y	-4	-6	-6	-6	-4	0	-6

The number -6 appeared four times in the bottom row.

The table was not symmetrical.

She had made some mistakes.

Five numbers in the table are incorrect.

Construct a new corrected table for the equation $y = x^2 - x - 6$

3   **a**   Construct a table for the graph of $y = -x^2 + 6x - 9$ for $0 \leqslant x \leqslant 6$

      b   Draw the graph of $y = -x^2 + 6x - 9$ for these values.

      c   Write down the coordinates of the maximum point.

      d   Work out the values of x for $y = -16$.

> **Hint**
>
> You may use an expanded table or a short table as in the first example in Learn 17.2.

4   **a**   Construct a table for the quadratic $y = x^2 + 1$ for $-3 \leqslant x \leqslant 3$

      b   Draw the graph of $y = x^2 + 1$ for these values.

      c   Write down the coordinates of the minimum point.

> **AQA Examiner's tip**
>
> Not all quadratics have three terms but the highest power will always be x^2.

5   **a**   Copy and complete this table for $y = 2x^2 - 7x + 5$

x	0	0.5	1	1.5	2	2.5	3
$2x^2$	0	0.5		4.5			18
$-7x$		-3.5	-7		-14		-21
$+5$	$+5$	$+5$	$+5$	$+5$	$+5$	$+5$	$+5$
y		2					2

      b   Draw the graph of $y = 2x^2 - 7x + 5$ for values of x from 0 to 3.

      c   Find the coordinates of the minimum point.

> **Hint**
>
> You should be able to see both from your graph and from the table at which x-value the minimum point will be.
>
> You might find it useful to add an extra column to your table. It can be placed at the end.

6

a Copy and complete this table for $y = (x - 4)(x + 2)$

x	−3	−2	−1	0	1	2	3	4	5
$(x - 4)$	−7				−3				
$(x + 2)$	−1				3				
y	+7				−9				

b Draw the graph of $y = (x - 4)(x + 2)$ for values of x from −3 to 5.

c Write down the coordinates of the minimum point.

d Use your graph to find the values of x when $y = 3$.

> **Hint**
>
> The equation is in bracket form.
> This will still give a quadratic.
> The two middle rows here have to be multiplied together.

7 A ball is thrown vertically upwards in the air.

After t seconds, its height above the ground, h metres, is given by the equation:

$$h = 39.2t - 4.9t^2$$

a Copy and complete the table for values of t from 0 to 8.

t	0	2	4	6	8
$39.2t$	0	78.4	156.8		313.6
$-4.9t^2$	0	−19.6		−176.4	
h		58.8	78.4		

b Draw the graph of $h = 39.2t - 4.9t^2$ for values of t from 0 to 8.

c At what time is the ball at its maximum height above where it was thrown from?

d What is this maximum value for h?

e At what times is the ball 40 metres above where it was thrown?

> **Hint**
>
> You will need to take h-values between 0 and 80 using 10 squares to equal 10. The graph will not be very accurate due to the values containing decimals. Just plot these as accurately as you can.

8 A flower bed is in the shape of a semicircle as shown. The quadratic equation $A = \frac{1}{2}\pi r^2$ gives the area for the flower bed in metres², where r is the radius.

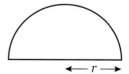

a Copy and complete the following table giving values of A to one decimal place.

r (metres)	0	1	2	3	4	5
$A = \frac{1}{2}\pi r^2$			6.3		25.1	39.3

b Draw the graph of $A = \frac{1}{2}\pi r^2$ using 2 cm to represent 1 m on the r-axis and 2 cm to represent 10 m² on the A-axis.

c Use your graph to estimate the area, A, of the flower bed when the radius is:

i 1.5 m   **ii** 3.8 m

d Use your graph to estimate the radius, r, of the flower bed when the area is:

i 30 m²   **ii** 16 m²

17   Assess _k!_

D

1   Which of the following are quadratic equations?

    a $y = x^2 + 3$ **b** $y = -2x^2$ **c** $y = x + 2$

2   **a**   Copy and complete the table of values for $y = x^2 + 3$

x	−3	−2	−1	0	1	2	3
y	12		4			7	12

    b   Draw the graph of $y = x^2 + 3$ for values of x from −3 to 3.

       (You will need y-values from 0 to 12.)

C

3   A rectangular enclosure has dimensions $(x + 2)$ metres by $(3 - x)$ metres.

$(x + 2)$ metres

$(3 - x)$ metres

The area of the enclosure is given by the formula

     $A = -x^2 + x + 6$

    a   Construct a table for the values of x from 1 to 5 inclusive.

    b   Draw the graph of $A = -x^2 + x + 6$ for values of x from 1 to 5 inclusive.

    c   What is the largest possible area this rectangle can have?

4   A designer was asked to include an arch of a particular shape in his plans for a building.

The equation he was given for the arch was $y = 4.5 - 0.5x^2$

This quadratic is symmetrical about the y-axis.

    a   Copy and complete this table for values of x from 0 to 3.

x	0	0.5	1	1.5	2	2.5	3
4.5	4.5	4.5		4.5		4.5	
$-0.5x^2$	0	−0.125			−2	−3.125	−4.5
y	4.5	4.375				1.375	

    b   Draw the graph of $y = 4.5 - 0.5x^2$ for values of x from −3 to 3.

       Use 2 cm to represent 1 unit on each axis. You will need to approximate your answers from the table.

       This will show you the whole arch.

    c   From this graph, estimate the value of y when $x = 1.2$

    d   Find the values of x for which $y = 4$

AQA Examination-style questions 🔴

1 **a** Explain why x^2 is never negative. *(1 mark)*

  b Copy the grid and draw the graph of $y = x^2$ for values of x from -3 to $+3$.

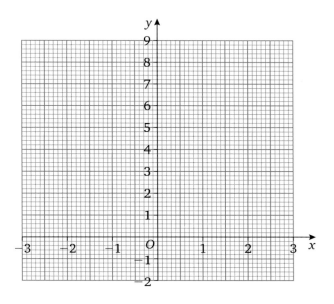

(2 marks)

AQA 2007

2 **a** Copy and complete the table of values for $y = x^2 - x - 5$

x	-2	-1	0	1	2	3	4
y	1		-5	-5	-3	1	

(2 marks)

  b Copy the grid below and draw the graph of $y = x^2 - x - 5$ for values of x from -2 to 4.

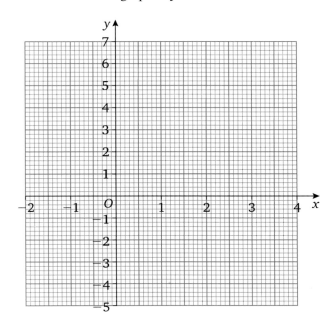

(2 marks)

AQA 2004

18 Pythagoras' theorem

Objectives

Examiners would normally expect students who get these grades to be able to:

C

use Pythagoras' theorem to find the third side of a right-angled triangle

use Pythagoras' theorem to prove that a triangle is right-angled.

You should already know:

✔ how to find squares and square roots using a calculator

✔ the properties of triangles and quadrilaterals.

Key terms

hypotenuse
Pythagoras' theorem

Did you know?

Pythagoras

Pythagoras lived in the 5th century BCE, and was one of the first Greek mathematical thinkers.

Pythagoras is known to students of mathematics because of the theorem that bears his name: 'The square on the hypotenuse is equal to the sum of the squares on the other two sides.'

The Egyptians knew that a triangle with sides 3, 4 and 5 has a 90° angle. They used a rope with 12 evenly spaced knots like this one, to make right angles.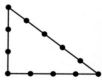

But they did not extend the idea to triangles with other dimensions.

Other people such as the Chinese and the Sumerians also already knew that it was generally true and used it in their measurements. However, it was Pythagoras who is said to have proved that it is always true.

Learn...   **18.1 Pythagoras' theorem**

In any right-angled triangle the longest side is always opposite the right angle.

This side is called the **hypotenuse**.

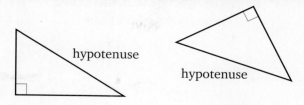

The diagram shows a right-angled triangle with sides of 3 cm, 4 cm and 5 cm.

Squares have been drawn on each side of the triangle and the area of each square is shown.

The area of the large square is equal to the sum of the areas of the two smaller squares.

$$25 = 9 + 16$$

This can be written as

$$5^2 = 3^2 + 4^2$$

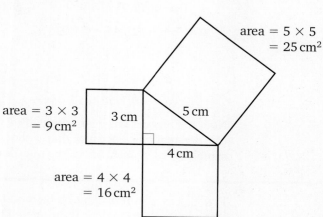

This relationship between the hypotenuse and the other two sides is true for any right-angled triangle. It is known as **Pythagoras' theorem**.

In general, in a right-angled triangle: $c^2 = a^2 + b^2$

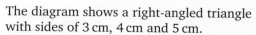

Not drawn accurately

Example: Calculate the length of the hypotenuse (labelled c) of this triangle.

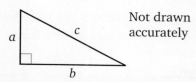

Not drawn accurately

Solution: Using Pythagoras' theorem,

$$c^2 = a^2 + b^2$$

$$c^2 = 5^2 + 7^2$$

$$= 25 + 49$$

$$= 74$$ Take the square root of each side.

$$c = \sqrt{74} = 8.6 \text{ cm (to 1 d.p.)}$$

AQA *Examiner's tip*

Check that your answers are sensible. If two sides are 5 cm and 7 cm then the third side cannot be 74 cm.

This will help you to remember to take the square root.

Example: Work out the length of side a.

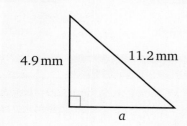

Not drawn accurately

Solution: This time you are trying to find a shorter side instead of the hypotenuse.

The hypotenuse is 11.2 cm.

Using Pythagoras' theorem,

$$c^2 = a^2 + b^2$$

$$c^2 - b^2 = a^2 \qquad \text{Subtract } b^2 \text{ from both sides.}$$

$$a^2 = c^2 - b^2 \qquad \text{Turn the equation round so that } a^2 \text{ is on the left.}$$

$$a^2 = 11.2^2 - 4.9^2 \qquad \text{Substitute in values for } c \text{ and } b.$$

$$a^2 = 125.44 - 24.01$$

$$a^2 = 101.43$$

$$a = \sqrt{101.43} \qquad \text{Take the square root of both sides.}$$

$$a = 10.1 \text{ cm (to 1 d.p.)}$$

Note the answer has been rounded to one decimal place because that is the same degree of accuracy as the question.

Pythagoras' theorem can also be used to test whether a triangle is right-angled by showing that the sides fit the theorem.

To test for a right angle, first square the longest side.

Then add the squares of the two shorter sides.

If your results are equal, the triangle contains a right angle.

Example: A triangle has sides of 8 cm, 15 cm and 17 cm.

Is the triangle right-angled?

Solution: $8^2 + 15^2 = 64 + 225 = 289$

$$17^2 = 289$$

Yes, the triangle is right-angled.

Right-angled triangles can be formed in other shapes. Some examples are shown below.

Wherever a right-angled triangle is formed, you can use Pythagoras' theorem to find the length of the third side.

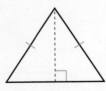

isosceles triangle

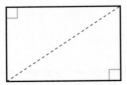

rectangle

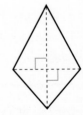

kite

Practise...   **18.1 Pythagoras' theorem**   G F E D C

C

1   Find the length of the hypotenuse in each of these triangles.

a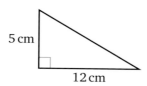
5 cm
12 cm

c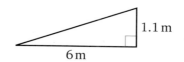
1.1 m
6 m

Not drawn accurately

b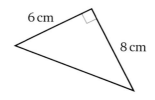
6 cm
8 cm

d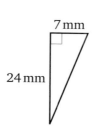
7 mm
24 mm

2   Find the length of the diagonal in each of these rectangles.

Give each answer correct to one decimal place.

a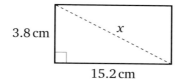
3.8 cm
x
15.2 cm

b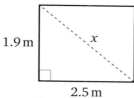
1.9 m
x
2.5 m

Not drawn accurately

c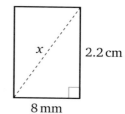
x
2.2 cm
8 mm

3   Find the length of the side marked x in each of these triangles.

Give each answer correct to one decimal place.

a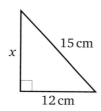
15 cm
x
12 cm

b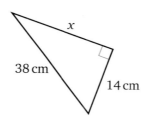
x
38 cm
14 cm

Not drawn accurately

c
x
4.2 m
13.7 m

4   A triangle has sides 24 cm, 26 cm and 10 cm.

Beth says that this is a right-angled triangle. Amy does not agree.

Who is correct?

Show working to justify your answer.

5   Find the length of the side marked x in each of these triangles.

Give each answer to one decimal place.

a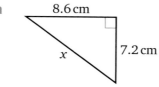
8.6 cm
x
7.2 cm

b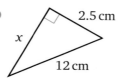
2.5 cm
x
12 cm

Not drawn accurately

c
4.5 cm
x
26 cm

d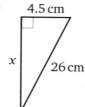
x
8 cm

C

6 Sarah and Ravi are working out the missing side in this triangle.
Sarah works out $7^2 + 22^2$ and says that $x = 23.1$ cm (to 1 d.p.)
Ravi works out $22^2 - 7^2$ and says that $x = 435$ cm
Both of these answers are incorrect.
Explain each person's mistake and work out the correct value of x.

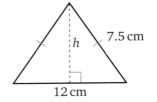

Not drawn accurately

7 An isosceles triangle has two sides of length 7.5 cm and one side of length 12 cm.
Calculate the height of the triangle.
Give your answer to one decimal place.

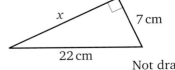

Not drawn accurately

8 Leon walks 1.5 km due north from his house.
He then turns and walks 2 km due east.
How far is he now from his house?

9 **a** A is the point $(2, 3)$ and B is the point $(5, 7)$ as shown in the diagram.

Use Pythagoras' theorem to find the distance between points A and B.

b Sketch a diagram and use Pythagoras' theorem to find the distance between each of the following sets of points.

  i $(3, 5)$ and $(7, 8)$

  ii $(0, 4)$ and $(5, 10)$

  iii $(-2, 3)$ and $(1, 7)$

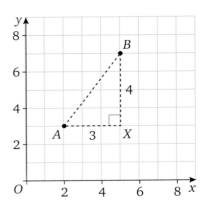

10 Find the length of the side marked x in each diagram.

a

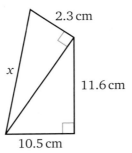

b

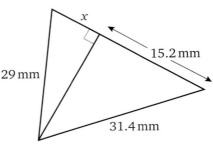

Not drawn accurately

11 A field is in the shape of a rectangle of length 45 metres and width 22 metres. A pipe runs diagonally from one corner of the field to the opposite corner.

How long is the pipe?

Not drawn accurately

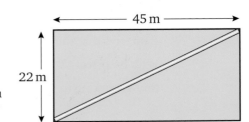

12 A ladder is 6.5 metres long.
The safety instructions say that for a ladder of this length:

- the maximum safe distance of the foot of the ladder from the wall is 1.7 metres

- the minimum safe distance of the foot of the ladder from the wall is 1.5 metres.

What is the maximum vertical height that the ladder can safely reach?
Give your answer to the nearest centimetre.

13   Cath has designed a pendant for a necklace in the shape of a kite.

Work out the length, x, of green line.

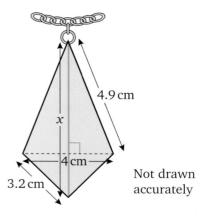

4.9 cm

x

4 cm

3.2 cm

Not drawn accurately

14   Pythagorean triples are sets of three integers that fit Pythagoras' theorem.

For example 3, 4, 5 is a Pythagorean triple as $3^2 + 4^2 = 5^2$. Similarly 5, 12, 13 and 7, 24, 25 are Pythagorean triples.

a   Find other Pythagorean triples.

b   What patterns can you see in the numbers?

c   See if you can find a rule to generate other triples.

18   Assess (k!)

1   Find the length of side x in each triangle.

Give each answer to one decimal place.

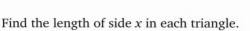

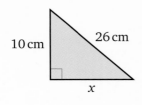

10 cm   26 cm x

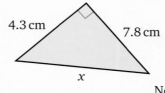

4.3 cm   7.8 cm x

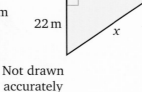

27 m   22 m x

Not drawn accurately

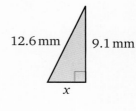

12.6 mm   9.1 mm x

2   Work out the perimeter of this quadrilateral.

> **Hint**
>
> Join B and D to form two right-angled triangles.

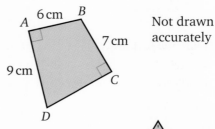

A 6 cm B

7 cm

9 cm

C

D

Not drawn accurately

3   An equilateral triangle has sides of length 15 cm.

Work out the perpendicular height of the triangle.

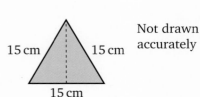

15 cm   15 cm

15 cm

Not drawn accurately

4   A field is in the shape of a rectangle of length 37 m and width 29 m.

A path runs diagonally from one corner of the field to the opposite corner.

What is the length of the path?

5   Rob walks from his house 3 km due south then 2 km due west.

How far is Rob now from his house?

C

6   A telegraph pole is kept in a vertical position by wires of length 10 metres and x metres that are fixed to the ground.

Calculate:

a   the height, h, of the telegraph pole

b   the length, x, of the second wire.

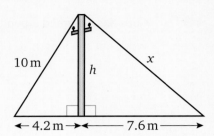

AQA Examination-style questions

1   The diagram shows a right-angled triangle. Calculate the length x.

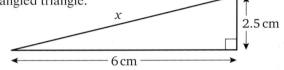

Not drawn accurately

(3 marks)

AQA 2009

2   A ladder of length 5 m rests against a wall. The foot of the ladder is 1.7 m from the base of the wall.

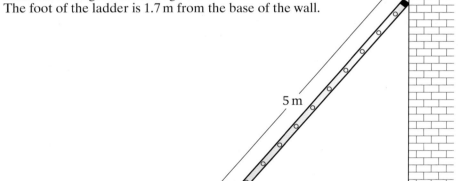

Not drawn accurately

How far up the wall does the ladder reach?

(3 marks)

AQA 2008

3   Calculate the length x.

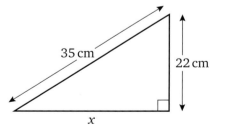

Not drawn accurately

(3 marks)

AQA 2007

You have covered the following topics:

Fractions and decimals

Angles

Working with symbols

Percentages and ratios

Perimeter and area

Equations

Properties of polygons

Coordinates and graphs

Reflections, rotations and translations

Formulae

Area and volume

Measures

Trial and improvement

Enlargements

Construction

Loci

Quadratics

Pythagoras' theorem

All these topics will be tested in this chapter and you will find a mixture of problem solving and functional questions. You won't always be told which bit of maths to use or what type a question is, so you will have to decide on the best method, just like in your exam.

Example: The diagram shows part of a golf course. There are nine holes.

The 7th hole is at *P*.

The 8th hole is at *Q*.

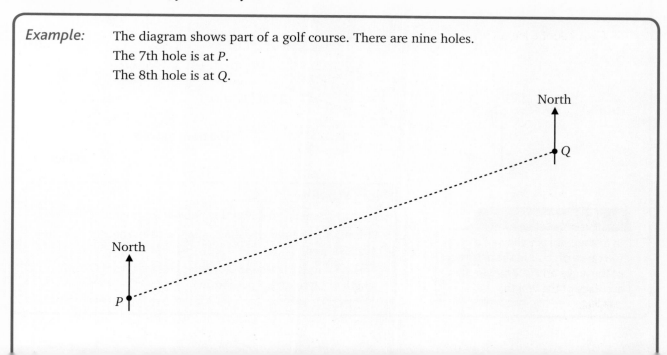

The diagram has been drawn using a scale of 1 : 200

a   What is the actual distance, in metres, between the 7th hole and the 8th hole?   *(3 marks)*

b   The 9th hole is 16 m from Q and on a bearing of 300° from Q.

Mark, with a cross, the position of the 9th hole on the diagram.   *(3 marks)*

Solution: **a**   The scale 1 : 200 means 1 cm on the diagram is equal to 200 cm on the actual golf course.

There are 100 cm in 1 metre.

200 ÷ 100 = 2

So 200 cm = 2 metres

So the scale 1 : 200 also means 1 cm on the diagram is equal to 2 metres on the actual golf course.

Start by measuring the length PQ on the diagram.

PQ = 12 cm

AQA Examiner's tip

In exams, there is a tolerance of ± 2 mm on measurements. So, if your measurement is between 11.8 and 12.2 cm you can get full marks.

However, always take care and check. Remember that measurements are **not** always a whole number of centimetres.

So the actual distance between the 7th hole and the 8th hole = 12 × 2 = **24 metres**

AQA Examiner's tip

You will lose 1 mark if you do not give your answer in metres.

Mark scheme

- 1 mark for measuring PQ to get 11.8 cm to 12.2 cm
- 1 mark for using the scale 11.8 to 12.2 × 200
- 1 mark for dividing by 100; this mark can be earned in two ways. Either for changing the scale to 1 cm : 2 m or by working out the answer in centimetres and then dividing by 100.

b   Convert 16 m to centimetres using the scale.

16 ÷ 2 = 8 cm

Draw a line from Q to represent the bearing 300°.
This is the blue line on the diagram.
Mark a cross on this line 8 cm from Q.
This is the position of the 9th hole.

9th hole ✕   16 m (8 cm)

North

Q

300°

24 m (12 cm)

North

P

Diagram to scale

AQA Examiner's tip

In exams, you are allowed a tolerance of ± 2° in the accuracy of the angle and ± 2 mm in the accuracy of the length.

Mark scheme

- 1 mark for working out that the required length from Q is 8 cm.
- 1 mark for a line showing the bearing, with an angle of 298° to 302° allowed.
- 1 mark for marking a cross along this line 8 cm from Q, with a length of 7.8 cm to 8.2 cm allowed.

Example: A company makes toy building blocks in the shape of cubes.

There are three sizes.

Some blocks have sides of 2 cm, some 4 cm and some 8 cm.

 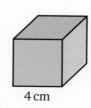 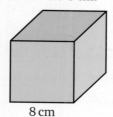

2 cm   4 cm   8 cm

The blocks are sold in boxes.

In a box there is:
- a bottom layer made up of the 8 cm blocks
- a middle layer made up of the 4 cm blocks
- a top layer made up of the 2 cm blocks.

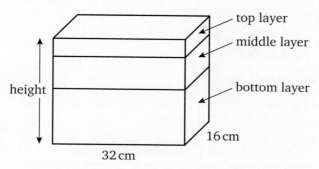

top layer

middle layer

height

bottom layer

16 cm

32 cm

a   What is the height of the box?   *(1 mark)*

b   Work out how many blocks are in the box.   *(3 marks)*

Solution: **a**   Think of an 8 cm block with a 4 cm block on top and a 2 cm block on top of that.

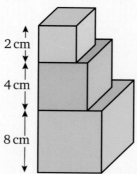

2 cm

4 cm

8 cm

AQA Examiner's tip

Don't forget to read everything carefully and use the diagrams to make sense of the question.

The height is   8 + 4 + 2 = 14 cm

This must be the height of the box.

The height of the box is 14 cm.

Mark scheme
- 1 mark for working out the height of the box correctly.

b   Start by thinking of the bottom layer. How many 8 cm blocks will there be?

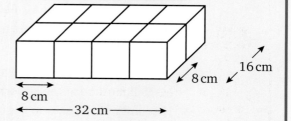

16 cm

8 cm

8 cm

32 cm

32 ÷ 8 = 4 means there are 4 blocks along the front

16 ÷ 8 = 2 means there are 2 rows

This will mean there are 4 x 2 = 8 of these blocks.

Repeat this working for the other layers.

For the 4 cm blocks in the middle layer.

32 ÷ 4 = 8 and 16 ÷ 4 = 4

So there are 8 of these blocks along the front and 4 rows.

There are 8 x 4 = 32 of these blocks.

For the 2 cm blocks in the top layer.

32 ÷ 2 = 16 and 16 ÷ 2 = 8

So there are 16 of these blocks along the front and 8 rows.

There are 16 x 8 = 128 of these blocks.

Now add up all the blocks.

Altogether there are 8 + 32 + 128 blocks

$$= 168$$

There are 168 blocks in the box.

> ### Mark scheme
> - 1 mark for knowing how to work out the number of blocks in one layer
> - 1 mark for working out the blocks in each of the three layers
> - 1 mark for the final correct answer

Consolidation

G

1   Stephen is making shapes using centimetre cubes.

1 cm

   a   What is the volume of this shape?

   b   Stephen has eight cubes altogether.
   He decides to make a large cube which is 3 cm high.
   How many more centimetre cubes does he need
   to complete his large cube?

2   A flight from Jersey to Southampton Airport on Easyfly Airways lasts 50 minutes.
A flight leaves Jersey at 09:40.

   a   What time will it arrive at Southampton Airport?

   b   Mr Bell has a meeting in Southampton.
   The meeting starts at 11:00.
   It takes 45 minutes for Mr Bell to get from the airport to the meeting.
   Does Mr Bell arrive on time for the meeting?
   Give a reason for your answer.

   c   The flight back to Jersey leaves Southampton at 17:30.
   Mr Bell must be at the airport at least 1 hour before the flight leaves.
   He allows 45 minutes to get from the meeting to the airport.
   What is the latest time Mr Bell must leave the meeting?

3 **a** Copy the diagram and add two more lines to it to make a polygon.

The dotted line must be a line of symmetry of your polygon.

b Here are the names of some polygons.

    hexagon   kite   octagon   pentagon   rhombus

Write down the one which is the name of the shape you made in part **a**.

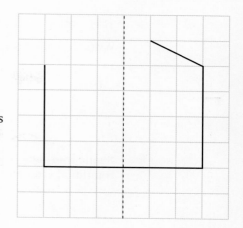

G

4 **a** Draw a circle with radius 5 centimetres.

b On your circle draw a chord of length 6 centimetres.

c Draw a diameter of your circle that is perpendicular to your chord.

5 The diagram shows a regular hexagon with all the possible diagonals.

a On a copy of the diagram shade all the triangles that are congruent to triangle *A*.

b How many triangles, congruent to triangle *B*, are in the diagram?

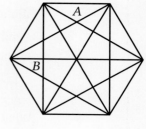

6 The point *A* is plotted on this grid.

a Write down the coordinates of *A*.

b The point *B* has coordinates (7, 6).
The point *C* has coordinates (3, 5).

   i Measure the length of the line *AB*.

   ii Measure angle *CAB*.

c Explain why triangle *ABC* is isosceles.

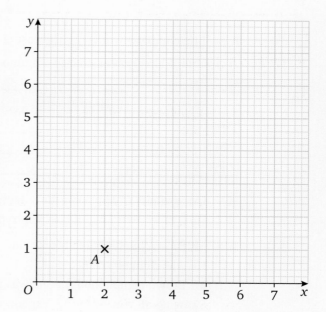

G
F

7 The diagram shows a sketch of a ladder against a wall.

a Draw the diagram accurately using a scale of 1 centimetre to 1 metre.

b Use your diagram to work out the length of the ladder. *ℓ*

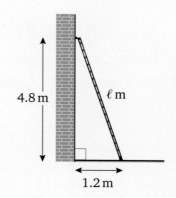

F

F

8   **a**   This symbol is the emblem of the Isle of Man.

Write down the order of rotational symmetry of the symbol.

b   Shade four more squares so that this shape has rotational symmetry of order 4 about its centre, *C*.

9   Clark is driving his car in France. His journey is 320 kilometres.

a   Clark says, '320 kilometres is not as far as 320 miles.'

Is Clark correct?

You must give a reason for your answer.

b   A milometer records the number of miles the car has travelled.

At the start of the journey the milometer reads

2	0	3	2	6

What will the milometer read at the end of the journey?

10   A car hire company works out the cost of hiring a car using the following formula:

cost of hire = fixed charge + number of days × daily rate

	Fixed charge	Daily rate
Small car	£10	£25
Medium size car	£15	£28
Large family car	£18	£30

a   Linda wants to hire a car for 3 days.

How much extra will it cost her to hire a medium size car than a small car?

b   Mr Brown hired a large family car.

He was charged £178.

Mr Brown complained that he was not charged the correct amount.

Showing all your working, explain why the company must have made a mistake.

11   *AB* and *CD* are straight lines.

AD and *DC* are perpendicular.

Work out the value of angles *x*, *y* and *z*.

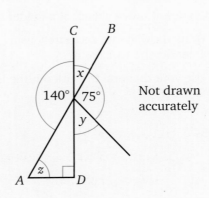

Not drawn accurately

12   One-half of a number is 15.

What is one-third of the number?

13   Two rectangles A and B can fit together to make a square.

The height of A is the same as the width of B.

The width of A is 3 cm.

The perimeter of A is 26 cm.

What is the perimeter of B?

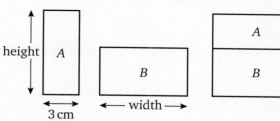

Not drawn accurately

14   **a**   Reflect this shape in the line $x = 3$

b   The whole shape is also symmetrical about a different line. Write down the equation of this line.

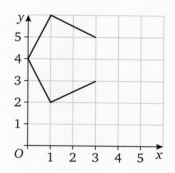

15   All of these shapes are drawn accurately.

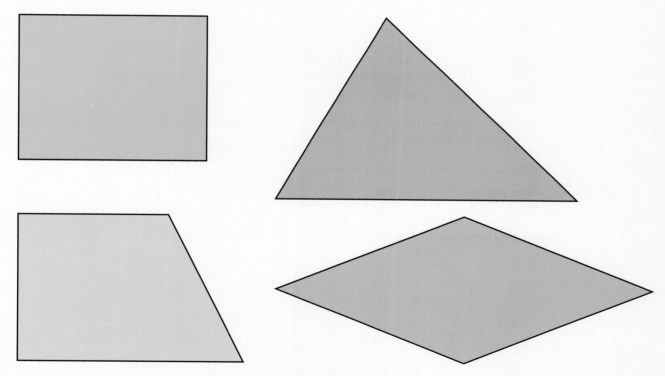

Show clearly that each shape has the same area.

16   There are some red counters, white counters and blue counters in a bag.

The total number of counters is 30.

$\frac{1}{3}$ of the counters are red.

$\frac{1}{2}$ of the counters are white.

Work out the fraction of the counters that are blue.

Give your answer in simplest form.

E

17   **a**   Complete the table below for the graph $y = 2x - 3$

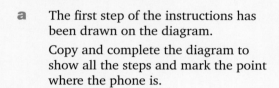

x	−1	0	1	2	3
y		−3			3

     b   Draw the graph $y = 2x - 3$

     c   Explain how the graph can be used to solve the equation $2x - 3 = 0$

18   Karla has hidden Jeremy's phone and mp3 player in a field.

She gives him these instructions to find his phone:

> From the gate, move 10 metres north.
>
> Then move 8 metres east.
>
> Then move 2 metres south.
>
> Your phone is here.

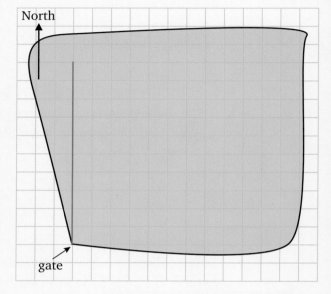

     a   The first step of the instructions has been drawn on the diagram.

         Copy and complete the diagram to show all the steps and mark the point where the phone is.

     b   Karla says:

> 'From where I hid the mp3 player I moved 3 metres south and 9 metres west to get back to the gate.'

         Copy and complete the following two instructions that will take Jeremy from where the phone is to where the mp3 player is.

         Move _____ metres _____

         Then move _____ metres _____

19   The graph shows the path of a cricket ball when it is thrown from a fielder to the wicketkeeper.

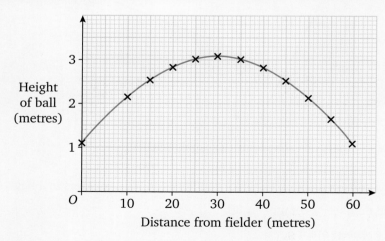

     a   How far is the wicketkeeper from the fielder?

     b   At what height does the wicketkeeper catch the ball?

     c   What is the greatest height above the ground that the fielder throws the ball?

20   **a**   Rachel draws two different quadrilaterals each containing **two** right angles.

   i   Her first quadrilateral has **one** line of symmetry.
      Draw **and** name this quadrilateral.

   ii   The second quadrilateral that Rachel draws has **no** lines of symmetry.
      Draw **and** name this quadrilateral.

   b   Rachel tries to draw a quadrilateral with exactly three right angles.
      Explain why she finds this impossible.

21   A hexagon with equal sides is cut along the dotted lines.
The dotted lines are parallel to the top and bottom
sides of the hexagon.
The four pieces A, B, C and D all have the same height.

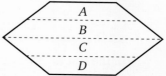

Not drawn
accurately

   a   What is the mathematical name of each of the four pieces?

   b   Draw a diagram to show how the four pieces can be put together to form a parallelogram.

   c   Shape D has an angle of 67.5° as shown below.

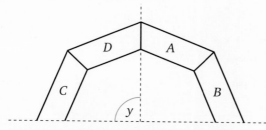

Not drawn
accurately

   Work out the value of x.

   d   The four pieces can be arranged symmetrically as shown below.
      The dotted lines are straight.

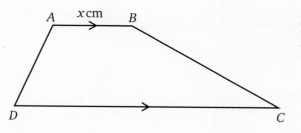

Not drawn
accurately

   Explain why angle y must be a right angle.

22   Simplify fully the following expressions.

   a $x + x + 2$ **c** $x + x \times 2$

   b $x \times x \times 2$ **d** $x \times x + 2$

23

Not drawn
accurately

$AB = x$ cm

BC is three times longer than AB.

CD is 2 cm longer than BC.

   a   Write down an expression in terms of x for the length CD.

   b   The perimeter of the shape is $2(4x + 3)$ cm.
      Work out the length of AD in terms of x.

D
C

24 In a test there are 10 questions.

If you attempt a question and get it right you score 5 marks.

If you attempt a question and get it wrong you lose 2 marks.

If you do **not** attempt a question you lose 10 marks.

a The table shows how three friends, Andrew, Bill and Clare, do in the test.

Name	Number of questions attempted	Number of questions correct
Andrew	10	6
Bill	9	8
Clare	8	7

In the test the mark for a Merit is 30.

The mark for a Pass is 20.

How well do the three friends do in the test?

You must show working to justify your answer.

b Tim takes the test.

He scores -7.

How many questions did Tim not attempt?

You must show working to justify your answer.

D

25 Here are details of Sudhir's bicycle journey.

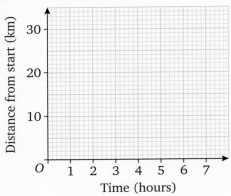

Stage 1: After the start he cycles at a speed of 12 km/h for $2\frac{1}{2}$ hours.

Stage 2: He stops for 30 minutes.

Stage 3: He cycles back towards the start for 1 hour travelling 10 km.

Stage 4: He stops for another 30 minutes.

Stage 5: He cycles back to the start at a speed of 8 km/h.

On a copy of this grid draw a distance–time graph to represent Sudhir's journey.

26 The diagram shows an irregular hexagon drawn on a centimetre square grid.

The dotted line shows an enlargement of one of the sides of the hexagon.

a Copy the diagram and complete the enlargement.

b Mark the centre of enlargement on the grid.

c Write down the scale factor of the enlargement.

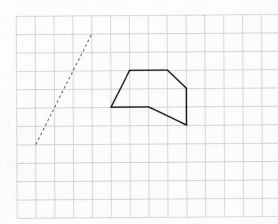

27 This solid shape is made from twelve cubes.
It is drawn on an isometric grid.
Draw three copies of this centimetre grid.

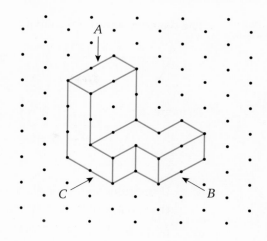

On your copies of the grid:

a draw the plan view from *A*

b draw the front elevation from *B*

c draw the side elevation from *C*.

28 Write down two different transformations
which map *A* onto *B*.

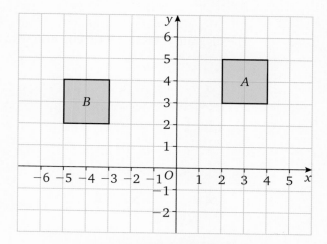

29 The volume of this cuboid is 480 cm³.

a Work out the width of the cuboid.

b The shape is cut in two, down the
dotted lines, and then glued together
to make a new cuboid.

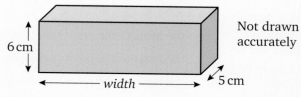

Not drawn
accurately

6 cm

← width →   5 cm

Not drawn accurately

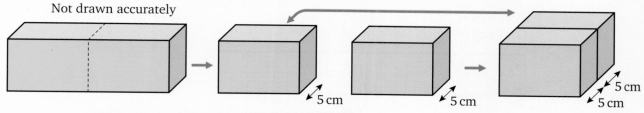

5 cm   5 cm   5 cm
                    5 cm

The surface area of the original cuboid is 412 cm².
Melissa says that the surface area of the new cuboid is the same as the original one.
Is Melissa correct?
You must show all your working to justify your answer.

30 The price of tickets for a boat trip to the Farne Islands is:

Adults   £12   Children   £8

x adults and *y* children go on a boat trip to the Farne Islands.
£*T* is the total price of their tickets.

a Write a formula for *T* in terms of *x* and *y*.

b On one boat trip the total price of the tickets is £672.
The number of children's tickets sold is 18.
How many adult tickets are sold on this trip?

D

31   The diagram shows a sketch of a triangular field, *ABC*.

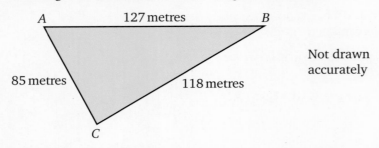

Not drawn accurately

a   Using ruler and compasses only, construct an accurate scale drawing of the triangle. Use a scale of 1 cm to 10 metres.

b   In the field there is a telegraph pole.
The telegraph pole is 44 metres from *A* and 63 metres from *C*.
How far is the telegraph pole from *B*?

32   The diagram shows a centimetre grid with the points *A*, *B*, *C* and *D* marked.

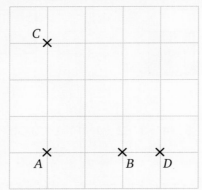

On a copy of the grid draw:

a   the locus of the point that is an equal distance from *A* and *B*

b   the locus of the point that is an equal distance from *A* and *C*

c   the locus of the point that is an equal distance from *C* and *D*.

D
C

33   **a**   Copy and complete the table for $y = x^2 + 5$

x	−3	−2	−1	0	1	2	3
y							

b   Draw the graph of $y = x^2 + 5$ for values of x from −3 to +3.

C

34   Plot the points $X(1, 1)$, $Y(10, 2)$ and $Z(6, 9)$ on a centimetre grid.

Using a ruler and compasses construct:

a   the perpendicular bisector of the line *XY*

b   the bisector of angle *XYZ*.

35   A rectangle has length x^2 and width $x + 3$.

The area of the rectangle is 40 cm².

Use trial and improvement to work out the value of *x*.

Give your answer to one decimal place.

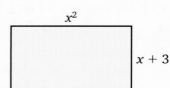

36   Mrs Sim has this photo frame.

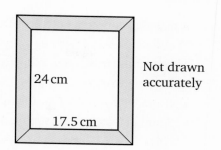

The photo frame is designed for a photograph of height 24 cm and width 17.5 cm.

Mrs Sim has a photograph of her daughter on her digital camera.

24 cm

Not drawn accurately

The print of this photograph has height 8 cm and width 5 cm.

17.5 cm

8 cm   Not drawn accurately

5 cm

She wants to print an enlargement of this photograph so that it fits the photo frame exactly.

Is this possible?

You must show working to justify your answer.

37

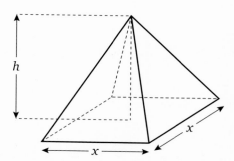

Not drawn accurately

h

x

x

The volume V of this square-based pyramid is given by the formula $V = \frac{1}{3}x^2h$ where x and h are measured in centimetres.

a   Work out the volume of the pyramid when $x = 6$ and $h = 8$

b   Aaron has been asked to make a square-based pyramid.

The height must be 2 cm bigger than the length of the base.

The volume of the pyramid must be 125 cm³.

He uses trial and improvement to work out the length of the base, x.

Complete his worksheet to find x to one decimal place.

Length of base, x cm	Height, h	x^2	Volume, cm³	
7	9	49	147	too big

38   A train usually travels along a section of track at 90 km/h for 6 minutes.

During repairs, the speed limit along this section is reduced to 20 km/h.

How much time will this add on to the journey?

39   Work out the length x in this triangle.

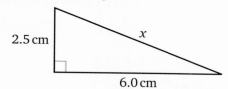

2.5 cm

x

6.0 cm

C

40 A machine cuts circular discs of diameter 5 cm from a sheet of rectangular plastic.

The dimensions of the sheet of plastic are 1.2 m by 0.8 m.

The machine leaves the following horizontal and vertical gaps:

- 2 mm between each disc
- 2 mm between the first row of discs and the top of the sheet
- 2 mm between the first column of discs and the left side of the sheet.

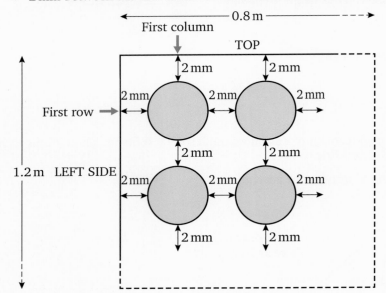

Not drawn accurately

a Show that 23 discs can be cut in each column.

b The unused part of a sheet is recycled.

What percentage of a sheet is recycled?

41 Use trial and improvement to find a solution to the equation $x^3 - 2x = 45$

The table shows the first trial.

x	$x^3 - 2x$	Conclusion
3	$3^3 - 2 \times 3 = 21$	Too small

Continue the table to find a solution to the equation.

Give your answer to one decimal place.

42 Adil and James are planning a mountain walk.

They find this rule to help them estimate how long the walk will take.

> **Estimating the time of a mountain walk**
>
> On a mountain walk it takes:
>
> 1 hour for every 3 miles travelled horizontally
>
> *plus*
>
> 1 hour for every 2000 feet climbed.
>
> Add 10 minutes of resting time for each hour you walk.

When planning their mountain walk, Adil and James estimate they will:

- travel 24 **kilometres** horizontally
- climb for 900 **metres**.

They plan to start their walk at 09:00.

At what time are they likely to complete their walk?

Hint

You will need to use these facts:
5 miles is approximately 8 km
3 metres is approximately 10 feet.

43   The diagram shows a scale drawing of Emma's bedroom.

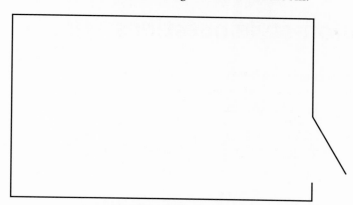

Scale : 1 cm represents 500 cm

Emma wants to paint the skirting boards in her bedroom.

Skirting boards are strips of wood at the bottom of the walls running around the edge of a room.

The skirting boards in her bedroom are 15 cm high.

Emma has a 1 litre tin of gloss paint.

According to the label, one litre of the paint covers between 10 and 12 square metres.

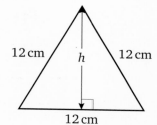

Does Emma have enough paint to paint the skirting boards with **two** coats of paint?

You must show working to justify your answer.

44   **a**   The sides of an equilateral triangle are 12 centimetres long.
Work out its height, h.

      b   A builder has a 2 metre long ladder.

He uses it to work on a wall and places the top of the ladder 1.8 metres from the ground.

To use the ladder safely the distance of the bottom of the ladder from the bottom of the wall, x, should be 0.45 metres.

Is the builder using the ladder safely?

12 cm h 12 cm

12 cm

AQA Examination-style questions

1 **a** Complete the table of values for $y = x^2 - 4x$

x	−1	0	1	2	3	4	5
y		0	−3	−4		0	5

(2 marks)

b On a suitable grid, draw the graph of $y = x^2 - 4x$ for values of x from −1 to 5. *(2 marks)*

c Use your graph to find the values of x *(2 marks)*

AQA 2008

2 Here are instructions for cooking a turkey.

Cook for 15 minutes at 220°C.

Reduce the oven temperature to 160°C and cook for 40 minutes per kilogram.

Kirsty is going to cook a 7 kilogram turkey.

She wants to take it out of the oven at 12. 45 pm.

At what time must she start to cook it? *(4 marks)*

AQA 2009

Glossary

acute angle – an angle between 0° and 90°

alternate angles – angles formed by parallel lines and a transversal that are on opposite sides of the transversal.

angle of rotation – the angle by which an object is rotated.

arc – a section of the circumference of a circle.

area – this is the amount of space that a shape covers.

axis (pl. **axes**) – the lines used to locate a point in the coordinates system; in two dimensions, the *x*-axis is horizontal and the *y*-axis is vertical. This system of Cartesian coordinates was devised by the French mathematician and philosopher René Descartes.

base – the lowest part of a 2-D or 3-D object (that is, the side that it stands on).

bearing – an angle that denotes a direction.

bisect – to divide into two equal parts.

bisector – a line that cuts either an angle or a line into two equal parts.

brackets – these show that the terms inside should be treated alike, for example,
$2 (3x + 5) = 2 \times 3x + 2 \times 5 = 6x + 10$

capacity – the amount of liquid a hollow container can hold, commonly measured in litres.

centre of enlargement – the point from which the enlargement is made.

centre of rotation – the fixed point around which an object is rotated.

chord – a straight line that joins any two points on the circumference, but does not pass through the centre.

circle – a 2-D shape made up of points that are all the same distance from a fixed point.

circumference – the distance all the way around a circle (that is, the perimeter of a circle).

compound measure – a measure formed from two or more measures. For example, $\text{speed} = \dfrac{\text{distance}}{\text{time}}$.

congruent – exactly the same size and shape; the shape might be rotated or flipped over.

construction (construct) – this is the process of drawing a diagram accurately with a 'straight edge' and compasses only.

conversion factor – the number by which you multiply or divide to change measurements from one unit to another.

coordinates – a system used to identify a point; an *x*-coordinate and a *y*-coordinate give the horizontal and vertical positions.

corresponding angles – angles in similar positions between parallel lines and a transversal.

cross-section – a cut parallel to a face of, and usually at right angles to the length of a prism.

cube – a solid with six identical square faces.

cuboid – a solid with six rectangular faces (two or four of the faces can be squares).

cylinder – a prism with a circle as a cross-sectional face.

decagon – a polygon with ten sides.

decimal place – the digits to the right of a decimal point in a number.

denominator – the bottom number of a fraction, indicating how many fractional parts the unit has been split into. For example, in the fraction $\frac{4}{7}$ the denominator is 7 (indicating that the unit has been split into 7 parts).

diagonal – a line joining two vertices (that are not next to each other).

diameter – the distance from one side of a circle to the other, through the centre. The diameter is double the radius.

dimension – the measurement between two points on the edge of a shape, for example, length.

edge – a line that joins two vertices of a solid. On a cube, such as a dice, an edge is the straight 'line' which is between each pair of faces.

elevation – this is the view of an object when viewed from the front or side; sometimes called front elevation (view of the front), or side elevation (view of a side).

enlargement – an enlargement changes the size of an object according to a certain scale factor.

equation – a statement showing that two expressions are equal, for example, $2y - 17 = 15$.

equidistant – to be equidistant from two points is to be the same distance from both points.

equilateral triangle – a triangle that has all three sides equal in length.

equivalent fractions – two or more fractions that have the same value. Equivalent fractions can be made by multiplying or dividing the numerator and denominator of any fraction by the same number.

expand – to remove brackets to create an equivalent expression (expanding is the opposite of factorising).

expression – a mathematical statement written in symbols, for example, $3x + 1$ or $x^2 + 2x$

exterior angle – the angle between one side of a polygon and the extension of the adjacent side.

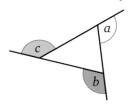

a, *b* and *c* are exterior angles

face – one of the flat surfaces of a solid. For example, a cube (such as a dice) has six flat faces.

factorise – to take common factors out of an equation or expression, often by the inclusion of brackets.

formula – a formula shows the relationship between two or more variables. For example, in a rectangle, area = length × width, or $A = lw$

hexagon – a polygon with six sides.

horizontal axis – in two dimensions, the x-axis is the horizontal axis.

hypotenuse – the longest side in a right-angled triangle. It is the side opposite the right angle.

image – a shape after it undergoes a transformation, for example, reflection, rotation, translation or enlargement.

improper fraction – or top-heavy fraction – a fraction in which the numerator is bigger than the denominator, for example, $\frac{13}{5}$ which is equal to the mixed number $2\frac{3}{5}$

integer – a whole number, positive, negative or zero, for example, -8, $+163$

interior angle – an angle inside a polygon.

a, b, c, d and e are interior angles

interior (or allied) angles – angles between two parallel lines and a transversal, which are on the same side of the transversal and between the parallel lines. For example, the angles marked c.

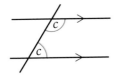

isosceles triangle – a triangle that has two sides equal in length.

like terms – $2x$ and $5x$ are like terms. xy and yx are like terms.

line of symmetry – a shape has reflection symmetry about a line through its centre if reflecting it in that line gives an identical-looking shape.

line of symmetry

locus (pl. loci) – a locus is the path followed by a moving point. It is also a set of points that meet a condition.

lowest common denominator – the lowest denominator that two or more fractions have in common. This is the least common multiple of the denominators of the fractions. For example, the fractions $\frac{2}{3}$, $\frac{1}{2}$ and $\frac{3}{4}$ have the lowest common denominator 12 because this is the least common multiple of 3, 2 and 4.

mass – the weight of an object, measured in tonnes (t), kilograms (kg), grams (g) and milligrams (mg).

mixed number – a fraction that has both an integer and a fraction part, for example, $1\frac{4}{7}$, which has 1 as the integer part and $\frac{4}{7}$ as the fraction part.

negative number – a number less than zero, expressed with a negative sign, for example, -5.3, -400

net – a net shows the faces and edges of an object. When the net is folded up it makes a 3-D object. For example, the net of a cube when folded up makes a cube.

nonagon – a polygon with nine sides.

numerator – the top number of a fraction, indicating how many parts there are in the fraction. For example, in the fraction $\frac{4}{7}$ the numerator is 4

object – a shape before it undergoes a transformation, for example, translation or enlargement.

obtuse angle – an angle between 90° and 180°

octagon – a polygon with eight sides.

operation – a rule for combining two numbers or variables, such as add, subtract, multiply or divide.

order of rotation – the number of ways a shape would fit on top of itself as it is rotated through 360° (shapes that are not symmetrical have rotation symmetry of order 1 because a rotation of 360° always produces an identical-looking shape).

origin – this is the starting point from which all measurements are taken.

parabola – the curved graph of a quadratic equation is called a parabola.

parallel – two lines that stay the same perpendicular distance apart.

pentagon – a polygon with five sides.

percentage – the number of parts per hundred, for example, 15% means '15 out of a hundred' or $\frac{15}{100}$.

perimeter – this is the distance all the way around a shape.

perpendicular – at right angles to; two lines at right angles to each other are perpendicular lines.

perpendicular height – the height of a shape that is 90° to the base.

plan – this is the view when an object is seen from above; sometimes called the plan view.

polygon – a closed two-dimensional shape made from straight lines.

positive number – a number greater than zero, sometimes expressed with a positive sign, for example, $+18.3$, 0.36

prism – a solid that has the same cross-section all the way through.

product – the result of multiplying numbers. For example, the product of 8 and 2 is 16.

proportion – if a class has 10 boys and 15 girls, the proportion of boys in the class is $\frac{10}{25}$ (which simplifies to $\frac{2}{5}$). The proportion of girls in the class is $\frac{15}{25}$ (which simplifies to $\frac{3}{5}$). A ratio compares one part with another; a proportion compares one part with the whole.

Pythagoras' theorem – in words 'the square on the hypotenuse is equal to the sum of the squares on the other two sides' or $c^2 = a^2 + b^2$

quadratic equation – a quadratic equation is one in which x^2 is its highest power. $y = x^2 + x - 2$ and $x^2 + x - 2 = 0$ are examples of quadratic equations.

quadratic expression – an expression containing terms where the highest power of the variable is 2

quadrilateral – a polygon with four sides.

radius (pl. radii) – this is the distance from the centre of a circle to a point on the circumference.

rate – the percentage at which interest is added.

ratio – a means of comparing numbers or quantities. It shows how much bigger one number or quantity is than another. If two numbers or quantities are in the ratio $1:2$, the second is always twice as big as the first. If two numbers or quantities are in the ratio $2:5$, for every 2 parts of the first there are 5 parts of the second.

reciprocal – the reciprocal of a number is 1 divided by that number. Any number multiplied by its reciprocal equals 1. For example, the reciprocal of 6 is $\frac{1}{6}$ because $6 \times \frac{1}{6} = 1$ and $1 \div 6 = \frac{1}{6}$

recurring decimal – a decimal whose digits after the point eventually form a repeating pattern. A dot over the digits indicates the repeating sequence, for example, $\frac{2}{7} = 0.\dot{2}8571\dot{4}$

reflection – a transformation involving a mirror line (or line of symmetry), in which the line from the shape to its image is perpendicular to the mirror line. To describe a reflection fully, you must describe the position or give the equation of its mirror line, for example, the triangle A is reflected in the mirror line $y = 1$ to give the image B.

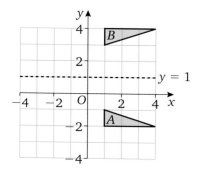

reflex angle – an angle between $180°$ and $360°$

regular – a shape that has all sides of equal length is regular.

right angle – an angle of exactly $90°$. It is always represented on a diagram by a small square.

right-angled triangle – a triangle with one right angle.

rotation – a transformation in which the shape is turned about a fixed point called the centre of rotation. To describe a rotation fully, you must give the centre, angle and direction (a positive angle is anticlockwise and a negative angle is clockwise). For example, the triangle A is rotated about the origin through $90°$ anticlockwise to give the image C.

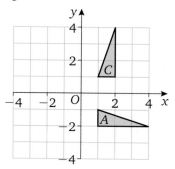

rounding – a number can be expressed in an approximate form rather than exactly; for example, it may be written to the nearest integer or to the nearest thousand. This process is called rounding. The number 36 754 rounded to the nearest thousand is 37 000

scale – the units used to measure along an axis.

scale factor – the scale factor of an enlargement is the ratio of the corresponding sides on an object and its image.

scalene triangle – a triangle with three sides of different lengths.

sector – this is the area between two radii and an arc.

segment – the area enclosed by a chord and an arc.

significant figures – numbers can be rounded to different numbers of significant figures. The number 30 597 when rounded to three significant figures is 30 600; when rounded to two significant figures it is 31 000. The closer a digit is to the beginning of a number the greater its significance in the number. Zeros are often used to maintain the place value of a number, but are significant figures if they are between other digits and, sometimes, if they are at the end of a number. The number 3.895 when rounded to three significant figures is 3.90: this zero is significant.

similar – shapes are similar (mathematically similar) if they have the same shape but different sizes, that is, one is an enlargement of the other.

simplify – to make simpler by collecting like terms.

solid – a three-dimensional shape.

solution – the value of the unknown quantity, for example, if the equation is $3y = 6$, the solution is $y = 2$

straight angle – an angle of $180°$

subject – the subject of the formula $P = 2(l + w)$ is P because the formula starts '$P = \ldots$'.

substitution (substitute) – in order to use a formula to work out the value of one of the variables, you replace the letters with numbers. This is called substitution.

sum – the result of adding numbers. For example, the sum of 8 and 2 is 10.

surface area – the exposed area of a solid object, often measured in square centimetres (cm^2) or square metres (m^2).

symbol – when there is an unknown, a symbol, such as a letter, is used to represent it.

symmetrical – a shape or graph is said to be symmetrical if reflecting it about a line through its centre gives an identical-looking shape.

tangent – a straight line outside a circle that touches the circle at only one point.

term – a number, variable or the product of a number and a variable(s), such as 3, x or $3x$

terminating decimal – a decimal that has a finite number of digits after the decimal point, for example, $\frac{1}{64} = 0.015625$

transformation – a transformation changes the position or size of an object. Examples of transformations include reflections, rotations, translations and enlargements.

translation – a transformation where every point moves the same distance in the same direction so that the object and the image are congruent.

transversal – a line that crosses two or more parallel lines.

trial and improvement – a method for solving algebraic equations by making an informed guess and then refining this to get closer and closer to the solution.

triangle – a polygon with three sides.

triangular prism – a prism with a triangular cross-section.

two-dimensional – a shape that has only two dimensions: length and width.

unit – a standard used in measuring. When a measurement is made units need to be chosen. Different units are appropriate for measuring items of differing size.

unitary method – a way of calculating quantities that are in proportion, for example, if 6 items cost £30 and you want to know the cost of 10 items, you can first find the cost of one item by dividing by 6, then find the cost of 10 by multiplying by 10.

unitary ratio – this is a ratio in the form $1 : n$ or $n : 1$. This form of ratio is helpful for comparison as it shows clearly how much of one quantity there is for one unit of the other.

unknown – the letter in an equation representing a quantity that is 'unknown'.

unlike terms – $2x$ and $5y$ are unlike terms. x and x^2 are unlike terms.

value – letters in a formula represent values. The given value of a letter is substituted into a formula to form an equation.

VAT (Value Added Tax) – this tax is added on to the price of some goods or services.

vector – a quantity with direction and magnitude (size). In this diagram, the arrow represents the direction and the length of the line represents the magnitude.

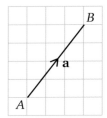

In print, this vector can be written as **AB** or **a.** In handwriting, this vector is usually written as $\overrightarrow{AB}$ or $\underline{a}$. The vector can also be described as a column vector:

where $\binom{x}{y}$ $\begin{array}{l} \leftarrow x \text{ is the horizontal displacement} \\ \leftarrow y \text{ is the vertical displacement} \end{array}$

vertex (pl. **vertices**) – the point where two or more edges meet.

vertical axis – in two dimensions, the y-axis is the vertical axis.

vertically opposite angles – the opposite angles formed when two lines cross.

volume – the amount of space a solid takes up, often measured in cubic centimetres (cm^3) or cubic metres (m^3).

Index

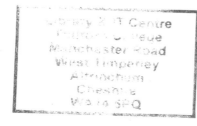